History of Astronomy S

Ancient Astronomy and Civilization

History of Astronomy Series

General Editor: A. G. Pacholczyk

ASTRONOMY BEFORE THE TELESCOPE
Nicholas T. Bobrovnikoff
This work presents a world-wide view of the development of astronomy in various countries with contrasting civilizations (including China and Maya) with emphasis on the interrelationship of its development with other sciences as well as with the political and cultural situations.
Volume One: The Earth-Moon System.
(Pachart History of astronomy Series, Volume 1)
ISBN 0-88126-201-3 $38.00

THE EXPANDING UNIVERSE:
LEMAITRE'S UNKNOWN MANUSCRIPT
M. Heller and O. Godart
A facsimile of an unpublished manuscript intended for the Catholic Encyclopedia of Japan, presents a totality of Lemaître's cosmology as well as his views on the existence of collapsed systems of stars and on the evolution of the Universe.
(Pachart History of Astronomy Series, Volume 2) $13.00

COSMOLOGY OF LEMAITRE
O. Godart and M. Heller
Written by two cosmologists, one of whom was a friend and a close collaborator of Lemaître, the book gives English speaking scientists and historians of science a correct picture of what contemporary science owes to Lemaître.
(Pachart History of Astronomy Series, Volume 3) $46.00

ANCIENT ASTRONOMY AND CIVILIZATION
Norriss S. Hetherington, illustrations by V. A. Mann
Both a text and a reader for science, philosophy, and history students, this book sketches the development of ancient science and philosophy, including interrelationships between science and society, and makes available in English translation generous excerpts from the works of early scholars, especially Plato, Aristotle, and Ptolemy.
(Pachart History of Astronomy Series, Volume 6)
ISBN 0-88126-225-0 $26.00

ANCIENT ASTRONOMY and CIVILIZATION

By Norriss S. Hetherington

Illustrations by V. A. Mann

PACHART PUBLISHING HOUSE

Library of Congress Catalog Card Number: 86-861671
International Standard Book Number: 0-88126-225-0

Pachart History of Astronomy Series Volume 6

Pachart Publishing House
1130 San Lucas Circle
P. O. Box 35549
Tucson, Arizona 85740

for Elizabeth

ACKNOWLEDGMENTS

The plan of Stonehenge is from STONEHENGE DECODED by Gerald Hawkins, in collaboration with John B. White. Reprinted by permission of Doubleday & Company, Inc.

The table of numbers of a Babylonian clay tablet is reproduced from Otto Neugebauer's *Exact Sciences in Antiquity, second edition*, with the permission of Dover Publications, Inc.

Greek commentaries on the chief pre-Socratic philosophers are reprinted from G. W. Kirk and J. E. Raven, *The Presocratic Philosophers*, with permission of the publisher and copyright holder, the Cambridge University Press.

Excerpts from Plato's *The Republic* are adapted from translations by G. H. Wells, by J. L. Davies and D. J. Vaughan, and by Francis Cornford.

Excerpts from Plato's *Timaeus* are reprinted from Francis Cornford's translation by permission of Humanities Press Inc., Atlantic Highlands, N. J. 07716.

Excerpts from Aristotle's *Physica* are from the translation by Richard Hope. Reprinted from ARISTOTLE'S PHYSICS translated by Richard Hope by permission of University of Nebraska Press. Copyright © 1961 by the University of Nebraska Press.

Excerpts from Aristotle's *De Caelo* are from the translation by J. L. Stocks. From the *Oxford Translation of Aristotle*, ed. W. D. Ross, vol. 2 (1930). Reprinted by permission of Oxford University Press.

Excerpts from Simplicius' commentaries on Aristotle's *De Caelo* and *Physica* are from Pierre Duhem's articles on the notion of physical theory from Plato to Galileo, collected and translated by E. Doland and C. Maschler in *To Save the Phenomena*. Reprinted by permission of the University of Chicago Press. © 1969 by The University of Chicago. All rights reserved.

The excerpt from Aristotle's *Metaphysics* is from the translation by Richard Hope. It is reprinted here by permission of Columbia University Press.

Excerpts from Simplicius' commentary on Aristotle's *De Caelo*, from Archimedes' *Sand-reckoner*, and from Aristarchus' *The Sizes and Distances of the Sun and Moon* are from *Aristarchus of Samos: The Ancient Copernicus* trans. Sir Thomas Heath (1913). Reprinted by permission of Oxford University Press.

Excerpts from Ptolemy's *Almagest* are from the translation by R. Catesby Taliaferro, published in volume sixteen of *Great Books of the Western World*. Reprinted by permission from *Great Books of the Western World*, copyright 1952 by Encyclopedia Britannica, Inc.

Contents

Preface

Too many bright students in the humanities have explained in tones more boastful than shameful that they are incapable of understanding science.

Too many students in the sciences with equally impressive intellectual abilities profess that the humanities are of no interest to them.

University requirements demanding a smattering of courses in a variety of fields dictate brief ventures into other students' preserves, but bridges across the culture gap are infrequent and little traveled. Interrelationships between the humanities and the sciences often are obscured by specialization within university faculties and the communication to students of artificial divisions into which knowledge has been cataloged.

Presented here is the development of a science and some of its cultural, historical, and intellectual interactions with the civilizations that created it. The skeptic, who doubts that interrelationships between the humanities and the sciences exist, is invited to read on.

The reader, skeptical or otherwise, will soon discover that a considerable proportion of this book consists of excerpts, in English translation, from the scientists' own writings. Studying literature or art, one would scarcely be content with summaries of novels or descriptions of paintings when the novel itself could be read or the painting viewed. The passages reproduced here are intended to facilitate an acquaintantship with scientific achievements, in their original form rather than in superficial summary.

What cannot be provided in a book is travel to foreign places, an essential part of an education. In the bibliographies at the end of each chapter are listed, along with a few recommendations for further reading, suggestions for films that might accompany this text in a classroom. Some of the films present technical details difficult to sketch on a blackboard; other films help convey a sense of earlier civilizations.

ANCIENT ASTRONOMY and CIVILIZATION

CHAPTER ONE

Megalithic Astronomy and Prehistory

Julius Caesar twice invaded Britain between 58 and 49 B.C. while he was conquering Gaul, and in 43 A.D. Rome finally conquered Britain, swelling the empire to its maximum extent. Marching inland, the Romans came upon the Plain of Salisbury and a gigantic pile of stones. Fifty-ton monsters stood upright to heights of over twenty feet, capped with other stones as if to form doorways. The Romans, though, had their own architectural marvels and were little impressed, either by Stonehenge or by the barbarian people inhabiting Britain.

Modern man is more impressed. Stonehenge now is recognized as a sophisticated astronomical observatory, and other megalithic or stone wonders also have been found in Britain. Evidence in stone contradicts the former belief that cultural, scientific, and technological innovations occurred only in Eastern civilizations and then were borrowed by the West. Previously pictured as howling barbarians, early Britons now are known to have enjoyed a civilization advanced enough to create Stonehenge. The discovery and study of early astronomical observatories furnishes important new evidence for a better understanding of past civilizations and for a rewriting of prehistory, of the period before written records were kept.

Legends grew up around Stonehenge. Roman writers mentioned it only infrequently, but during the Dark Ages speculation arose that Merlin, King Arthur's magician, had created Stonehenge as a monument to slain warriors. Already there was some appreciation of the technological accomplishment in moving stones of immense size and erecting them at the site. Magical powers were the explanation first offered, though, rather than the skills of an advanced civilization.

When Stonehenge finally was recognized, in the seventeenth century, as the handiwork of a mighty civilization, it was to the Romans that credit first was given. And if not to the Romans, next to the Druids. They may have risen in Britain around the fifth century B.C. Caesar later described them as heading a priestly cult and including public sacrifice among their rituals. Modern evidence, however, places the construction of Stonehenge earlier than either the Romans or the Druids.

Not until 1740 was it suggested that Stonehenge might be associated with the science of astronomy. Then a supporter of the Druid theory noted that the principal avenue of Stonehenge points to midsummer sunrise. A few other writers commented upon the possible astronomical significance of Stonehenge. A century and a half would pass, however, before the possibility of using the science of astronomy to assist in the reconstruction of history was recognized.

In 1890 J. Norman Lockyer, a British astronomer on vacation in Greece, noticed that in some instances the foundations of Greek temples had been changed after their original construction, apparently to realign the temples in slightly different directions. Lockyer also remembered that a properly constructed English church has its east window facing sunrise on the day of its patron saint. Could there be, Lockyer asked himself, a connection between astronomical phenomena and successive alignments of Greek temples?

From a companion Lockyer borrowed a pocket compass to make measurements. Back in Britain, Lockyer expanded his study to include Egyptian temples, excavated and described earlier in the century by German and French expeditions engaged in an honorable rivalry less destructive than war.

The Roman architect Vitruvius had written in the first century B.C. that Egyptian temples were built facing the Nile River. It required but a quick glance at temple plans to see how far from the truth was Vitruvius' statement. The Egyptians seemingly had suffered from symmetriphobia.

Lockyer, however, encouraged by the example of English churches, hoped to find some order. He set out to classify temples according to the direction of their main axes. The more temples that could be grouped together, the more likely was the discovery of a general law governing their structure.

Lockyer had an idea what he might find. Among astronomical objects commanding the alignment of temples, the sun is an obvious possibility. It's motion is connected with the important agricultural seasons and was well known to ancient man.

The seasons follow from the inclination of the earth's axis at an angle of 23.5 degrees to the plane of the earth's annual revolution around the sun. The northern hemisphere more fully faces the sun in summer, and is tilted away in winter. The longest day of the year, summer solstice, occurs at midsummer, when the sun is highest in the sky. Winter solstice is the shortest day of the year, with the sun lowest on the horizon.

Summer solstice in Egypt coincided with the rise of the Nile River, and it marked the beginning of the ancient Egyptian year. At Thebes at summer solstice, the sun rises on the horizon 26 degrees north of due east, and sets 26 degrees north of due west. At winter solstice, the sun rises 26 degrees south of due east and sets 26 degrees south of due west. With the solstice positions in mind, Lockyer examined data on the alignment of Egyptian temples.

The Seasons

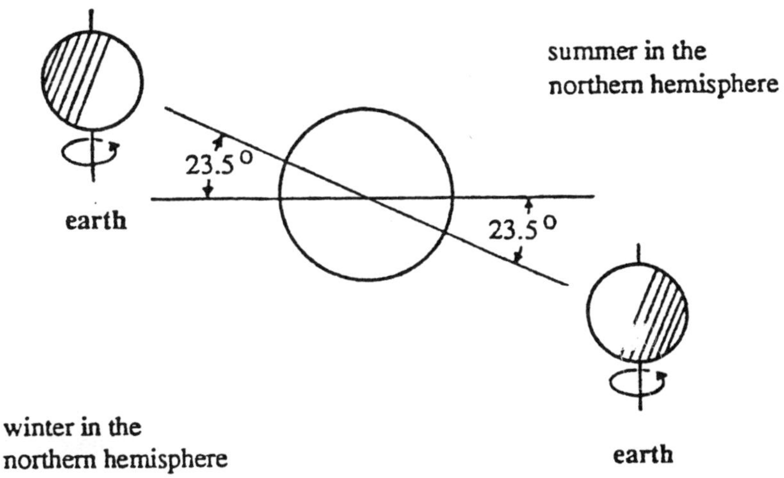

The fertile Nile Valley was occupied perhaps as early as 5000 B.C. Several small kingdoms were unified under the pharaohs, whose Old Kingdom lasted from about 3100 B.C. to 2200 B.C. Egypt then suffered a period of anarchy, followed by the Middle Kingdom, from about 2050 to 1750 B.C. This kingdom was established by princes at Thebes, some way up river from the mouth of the Nile.

Temples at Thebes, including Karnak and Luxor, furnished evidence for Lockyer's conclusion that Egyptian temples were oriented toward the sun at some particular time of the year. He found temples at Karnak oriented 26.5 to 27.5 degrees north of west, pointing very nearly to sunset at summer solstice. The main axis of another temple at Karnak pointed 26.5 degrees south of east, to sunrise at winter solstice. Other temples had similar alignments.

Lockyer also found a second solar alignment. On the pyramid plains at Gizeh and Memphis many temples were oriented east-west, pointing to sunrise and sunset at spring and fall equinox.

The Equinoxes

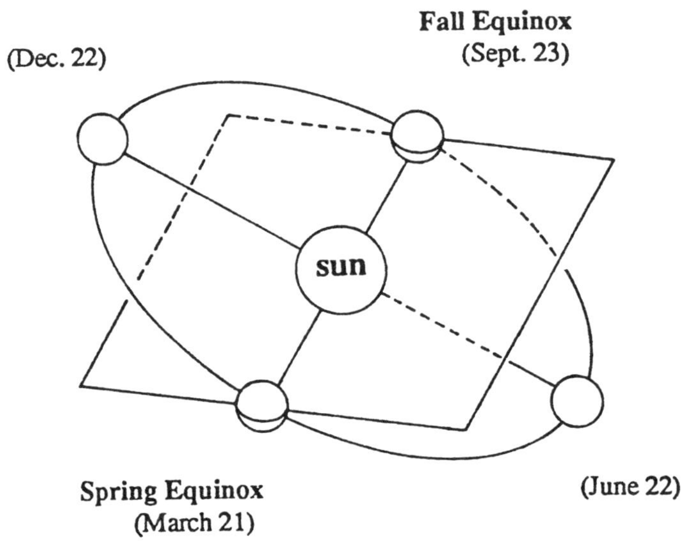

The plane of the earth's orbit around the sun is inclined 23.5 degrees to the plane of the earth's equator.

Because their temples were aligned toward the sun, Lockyer believed that the Egyptians had worshipped the sun. His intellectual leap from temple alignments to the religious and philosophical beliefs of ancient Egyptians was not, however, necessarily valid. His observation of the alignment of English churches had not carried him away on a similar flight of fancy, nor had the observation that Saint Peters in Rome is aligned exactly east-west and at the spring equinox the rays of the rising sun pass through opened outer and inner doors to illuminate the high altar.

Lockyer thought it also likely that some temples might have had their main axes aligned toward the position on the horizon where a star rose or set at a critical time. If stars always rose and set at the same point, explorers could travel to ancient temples, look along their axes, and observe what stars rose or set on propitious days. The position of a star, however, changes about 47 degrees in thirteen thousand years. Thus a temple aligned with the rising or setting of a star at one time in history will slowly lose its alignment. (The sun's position also changes with time, but only by one degree in seven thousand years, or 0.02 degrees per year.)

Frequent changes in the direction of the axes of some temples supported Lockyer's hypothesis that the temples were aligned towards stars. Even more convincing was the construction of new temples on old temples when, apparently, it was too difficult to realign the axis.

To determine a stellar alignment, some idea of the date of construction of the temple is required. Conversely, if inscriptions on the temple reveal which star the temple was lined up with, the date of construction can be calculated.

At Denderah, excavated by a French expedition in 1798, the temple of Isis bears an inscription stating that on New Years Day the star Sirius shines into the temple, mixing her light with that of her father the sun on the horizon. Lockyer calculated that the temple had pointed to Sirius in about 700 B.C., a date in agreement with that derived earlier from the positions of stars displayed in a zodiacal medallion found at the temple.

Encouraged, Lockyer went on to find a series of three temples, built about 2525, 1250, and 900 B.C., their alignments slowly changing with precisely the speed necessary to keep up with a star. After this discovery, evidence of stellar alignments of Egyptian temples flowed in in such abundance that Lockyer had to present his results in tables.

It is not uncommon to overlook familiar aspects of one's own culture and background, and then to be impressed upon seeing a similar object or condition in a new setting. This is what happened to Lockyer on his trip to the Near East, where he first became conscious of temple alignments. Back in England with heightened sensitivity, Lockyer turned his attention to Stonehenge. As a sequel to his work on Egyptian and Greek temples, he calculated that Stonehenge had faced sunrise at summer solstice about 1680 B.C.

Within a year Lockyer's date for the construction of Stonehenge was roughly confirmed. Archaeologists excavating, in order to re-erect a large and leaning stone likely to fall, found many flint instruments, including axes and hammers. The instruments were dated to about 1800 B.C.

Archaeologists, however, were not impressed by Lockyer's work. There was a feeling that the astronomer should stick to astronomy. Lockyer's brief examination of Stonehenge was insufficient to sway his critics. Not until the 1960s, with assistance from the new computer technology, could an astronomer overwhelm the skeptics with a mass of probable alignments.

Gerald Hawkins grew up in England only vaguely interested in Stonehenge, thinking that the Druids had built it and had probably had human sacrifices there. In 1953 he worked near Stonehenge, at a missile testing base, and often visited the pile of stones. Reading up on it, Hawkins found it remarkable that the monument was aligned to midsummer sunrise, even more remarkable than the nearly-riotous celebrations held there each summer by modern merrymakers. Fifteen military policemen were required to restore order in the summer of 1966.

Hawkins soon left for America, having done little more than marvel at Stonehenge. His interest in the ruins was renewed only when he began writing a book on astronomy.

Two stones are sufficient to mark the position of midsummer sunrise, but there are some 165 stones, holes, and mounds at Stonehenge. Possibly, thought Hawkins, all of the stones might have been placed to mark various astronomical alignments

He returned to England to make the measurements necessary to test his hypothesis. Hawkins chose from the 165 recognized positions 120 pairs of points. He drew lines through the points and determined to what part of the sky the lines pointed. This task would have taken a human about four months of hard work, but was done by a computer in minutes.

Many of the lines pointed to particular regions of the sky. Twelve were associated with the sun, as Hawkins had expected. What he had not anticipated were some dozen lunar alignments.

Given the large number of points and lines drawn between pairs of points, it was inevitable that some of the lines should point by sheer chance to significant astronomical phenomena. Hawkins' critics were quick to note this fact. But the number of correlations seemed, at least to Hawkins, too great to be coincidental.

It was also difficult to believe that a primitive people had built the sophisticated astronomical observatory discovered by Hawkins in a mass of computer printout. One scientist objected that "on the basis of ambiguous or implausible evidence and careless or suspicious research we have been asked to believe that certain celebrated barbarians, being unable to read or write, made observations and computed eclipses in the strangest ways with great huge rocks and hills."

Plan of Stonehenge

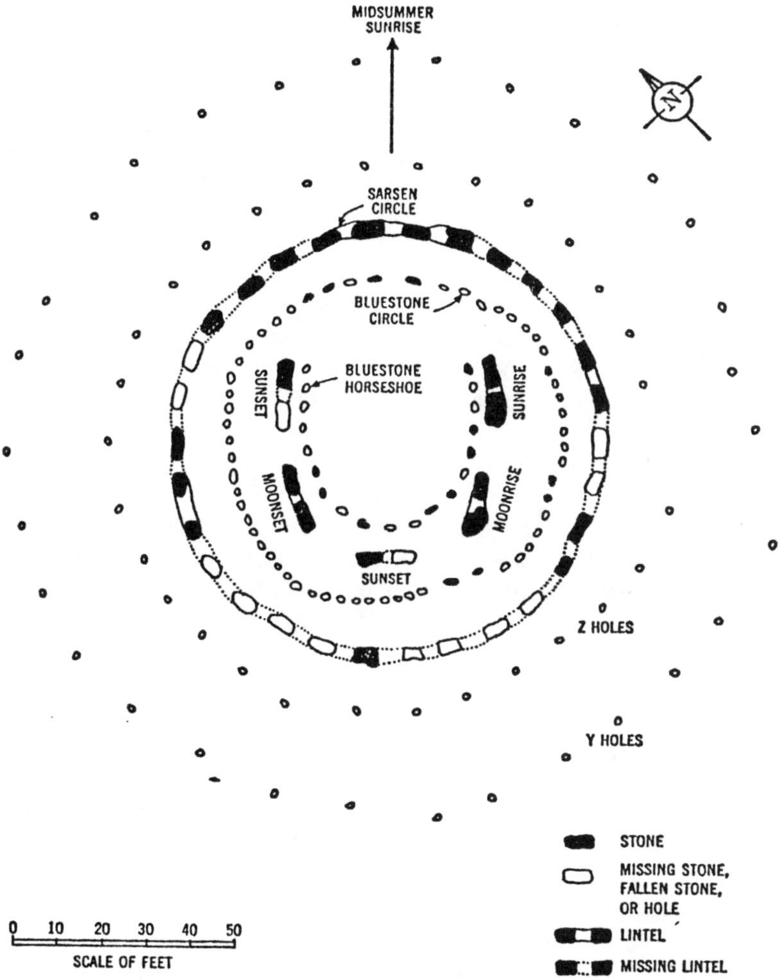

Further research, though, on stone circles in Britain, including circles such as Stonehenge, menhirs (single stones), stones in single or multiple parallel rows (avenues), irregular circles (cromlechs), three upright stones with one side open (coves), and flat table stones resting on two or more upright stones (dolmens) has done much to turn the tide of critical opinion. A retired professor of engineering, Alexander Thom, and members of his family have made detailed surveys of Stonehenge and other megalithic sites in Britain and have found many stones at many sites aligned toward astronomical objects. The Thoms' discoveries make up an overwhelming mass of detailed evidence and go far toward forcing acceptance of a revolutionary theory contradicting previous belief.

The Thoms have also shown that stones in parallel rows were set out with remarkable accuracy. The precision is more than could have been attained by using ropes. It may have been accomplished with measuring rods of oak or whalebone, a sort of megalithic yard stick. Somehow ancient Britons attained accuracy in their measurements comparable to that of a trained surveyor using modern equipment. Accurate, consistent measures for hundreds of years over a large region of Britain, as far north as the Orkney Islands, and even across the channel in France, suggest that the society was unusually well organized and administered.

A bit of doubt may remain over whether Thom really has discovered the megalithic yard or whether he has created a megalithic myth. It just may have been possible to have set the stones accurately by pacing off their positions. Even skeptics, though, concede that the Thoms have uncovered a great amount of important information. The difficult terrain and vegetation at Carnac in South Brittany, which seems to have been the megalithic metropolis of Europe, have made the Thoms' operation in prehistoric research one of the most extensive undertaken in Europe in the twentieth century.

Thom, Hawkins, and others have demonstrated that early Britons had a knowledge of mathematics and possessed extensive astronomical data. No evidence, however, has yet been found among the surviving archaeological material to show that these people kept any written records. How they retained their data and transmitted it from one generation to the next is difficult to imagine.

The discovery of early astronomical accomplishments has shown that the traditional prehistory of Europe is wrong. It must be revised. The new and growing discipline of astro-archaeology will contribute to the revision.

The successes at Stonehenge and at other sites in Britain have led to a flurry of investigations of possible astronomical alignments in other regions. In the dry foothills of the Andes in Peru, pebbles were scraped away long ago, leaving a huge geometrical pattern of lines. The possibility of astronomical alignments is attractive, but a recent study found none.

There do appear to be astronomical alignments of Mayan buildings on Mexico's Yucatan peninsula dating from about 200 to 900 A.D. In this case there are also written records, and they furnish much more extensive evidence of a flourishing culture and a high level of scientific knowledge. Indian pueblos in the American Southwest and Indian mounds in the Midwest also are beginning to yield evidence of early astronomical knowledge.

Almost certainly additional sites will be discovered. Cooperation between the previously separate academic disciplines of science and history will advance future studies. Each subject will be enriched by contributions from the other.

Bibliography

J. Norman Lockyer's work on Egyptian temples is reported in *The Dawn of Astronomy*. For Lockyer's investigations at Stonehenge, see his *Stonehenge and other British Monuments Astronomically Considered.*

Gerald Hawkins' account of his investigation, including diagrams and photographs, is published in *Stonehenge Decoded.*

The plan of Stonehenge is from STONEHENGE DECODED by Gerald Hawkins, in collaboration with John B. White. Copyright © 1965 by Gerald Hawkins and John B. White. Reprinted by permission of Doubleday & Company, Inc.

Alexander Thom's work is published in *Megalithic Sites in Britain*, in *Megalithic Lunar Observatories*, and in a series of articles in the *Journal for the History of Astronomy.*

The proceedings of a symposium on the place of astronomy in the ancient world were published in the *Philosophical Transactions of the Royal Society of London, volume 276* (1974), pages 1-276.

Later studies are reported in the journal *Archaeoastronomy*, which was born in 1979, an offspring of the *Journal for the History of Astronomy.*

Not to be missed is the film *Mystery of Stonehenge* (57 minutes), featuring Gerald Hawkins.

Alexander Thom and several megalithic sites, including Stonehenge, appear in *Cracking the Stone-Age Code* (52 minutes).

Also good is *The Sun Dagger* (59 minutes), featuring a celestial calendar built by Anasazi Indians a thousand years ago in Chaco Canyon, New Mexico.

The film *Earth and Its Seasons* (11 minutes) illustrates the equinoxes and the solstices.

Exploring the Unwritten Past (22 minutes) discusses stratification, carbon-14 dating, and evidence of ancient man in Britain.

Films on Egypt include *Ancient Egypt: The Sun and the River* (58 minutes), a history of Egypt; *Egypt: Gift of the Nile* (29 minutes), on the old, middle, and new kingdoms; *Ancient Egypt - revised edition* (11 minutes), on Egypt's contributions to civilization, including mathematics; and *Nile River Valley and the People of the Lower River* (17 minutes), with scenes of temples and pyramids and a discussion of the importance of the river.

For a brief pictorial sketch of Mayan civilization, see *Ancient Maya Indians of Central America* (9 minutes) in the *Ancient American Indian Civilization Series.*

CHAPTER TWO

Babylonian Clay Tablets

Recent research results in Northern-European prehistory are impressive. More, however, is known of Oriental civilizations in the millennium before Christ. And in the Orient, more so than in Northern Europe, the origins of Western civilization are found. Our roots are in Greek culture, which owed much to Oriental civilizations.

Communications between civilizations can constitute an important aspect of history. Sometimes cultural exchanges can be inferred from similarities between religious and philosophical ideas. Such speculation, however, is fraught with uncertainty. More precise evidence of cultural intercourse occasionally is found in the transmission of complicated astronomical methods and exact numerical values.

Civilizations emerged in the great river valleys of the Near East some six thousand years ago, in the Nile River Valley of Egypt and in the Tigris-Euphrates River Valley of Mesopotamia, now Iraq. Egyptian technology and practical science were highly advanced, with a useful system of numbers and a practical calendar. Egyptian temple alignments testify to a good working knowledge of celestial positions. Pure science, however, seems to have held little interest for the Egyptians. More interesting developments in theoretical science have been found in Mesopotamia.

Differences between these two ancient civilizations, Egypt and Mesopotamia, may be partly explained by geographic factors. Egypt was isolated and secure from invasion. Mesopotamia, in contrast, was vulnerable to raiders. The Mesopotamians' greater interest in predicting the future possibly followed from their greater insecurity, and may have lent a greater utilitarian justification to astronomical studies in Mesopotamia than in Egypt. Effort and creative thought in general are encouraged by insecurity, so some historians speculate.

Also, Egypt did not develop urban communities on the scale of those in Mesopotamia. And urban communities, whether the result or the cause of feelings of insecurity, do seem to spawn intellectual endeavor more often than do rural communities.

Motivation is a subjective and elusive subject, tempting to dismiss. But it is not unimportant. In addition to methods and achievements, an appraisal of Mesopotamian science examines motives too. Historians of science ask whether the inquiring spirit and theoretical orientation of modern science appeared in Mesopotamian civilization, or only later, with the rise of Greek civilization.

Mesopotamia is fertile ground for archaeological investigations. The people inscribed documents on clay tablets, an indelible record for others who came after them. Fragments of ancient pottery well illustrate the virtual indestructibility of clay.

For a long time, though, searchers in quest of the ancient civilizations of Babylonia and Assyria ignored the best sources of information. Mohammedans may have avoided the clay tablets because the Koran said they were bricks inscribed by demons and baked in Hell. Christians may have thought it blasphemy to revive knowledge of the Assyrians, whom the Lord had determined to destroy along with all record of their wickedness.

Not until the nineteenth century were clay tablets recovered in large numbers. Then a French official in Mesopotamia grew curious about some large pieces of sculpture, began excavations, and discovered the ancient city of Nineveh, the capital of the Assyrian Empire. Next he found remains of an earlier Sumerian Empire, which had controlled the lower part of the Tigris-Euphrates Valley from around 3000 to 2000 B.C.

The Sumerian rule was followed by that of the Amorites, who migrated from the Arabian Desert. Hammurabi, their greatest leader, ruled the city-state of Babylon from about 1792 to 1750 B.C. He is most famous for his code of laws.

The next rulers, the Assyrians, also came from the Arabian Desert. They settled in the upper part of the Tigris-Euphrates Valley. Pursuing a deliberate policy of terrorism - covering the walls of conquered cities with the skins of the captured populace - the Assyrians gained wide power in the ninth century B.C. They maintained control over a vast region until a series of revolts in the seventh century ended with the destruction of Nineveh in 612 B.C.

The successful rebels, the Chaldeans, ruled for less than a century. The Persians succeeded them, capturing Babylon in 539 B.C. Alexander the Great conquered the Persian Empire in 330 B.C., and his followers established the Seleucid rule of the last three centuries B.C.

From this period come most of the astronomical tablets recovered. The majority of mathematical texts recovered are from the earlier period of Hammurabi.

Map of Mesopotamia

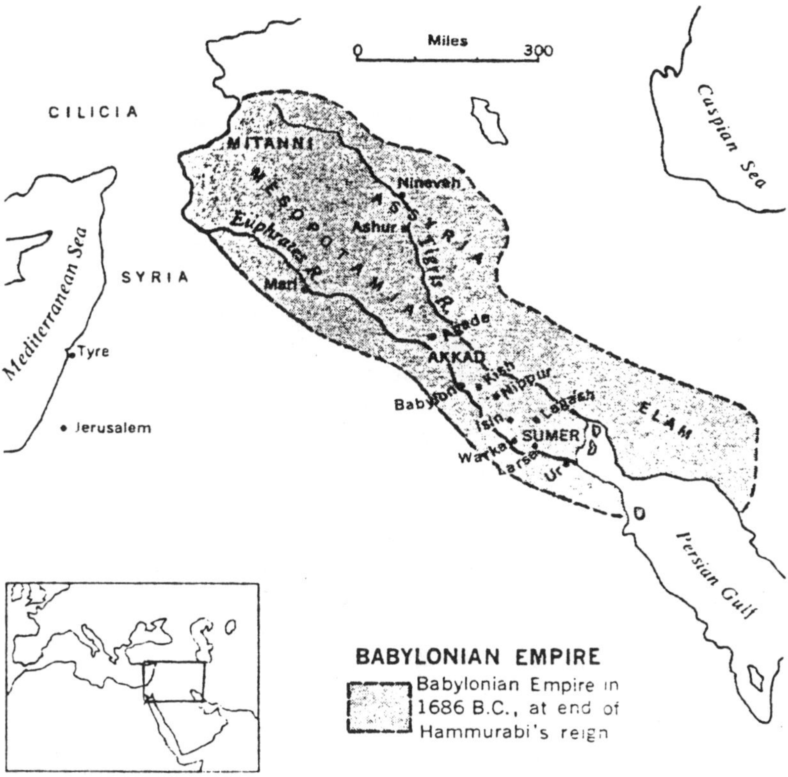

Shortly after the discovery of the ruins of Nineveh, the library of Ashurbanipal, an Assyrian king who died in 626 B.C., was uncovered. Some 22,000 clay tablets from that library now are stored in the British Museum. Another 50,000 clay tablets from the temple library of Nippur, written between 3000 and 450 B.C., form another valuable collection. There may be as many as half a million Babylonian clay tablets scattered among the museums of the world. Many more lay buried still in Mesopotamia.

An expedition to uncover clay tablets is more glamorous than the task of deciphering the tablets. Consequently, the number of tablets studied is less than the number unearthed. Enough texts have been studied, though, to provide the beginnings of an appreciation of Babylonian astronomy. And there is evidence for the transmission of numerical parameters of Babylonian astronomy to the Greek and Roman world.

According to Greek tradition, Babylonian and Chaldean astrology and number mysticism were the dominant aspects of Babylonian science. This appraisal has been revised in the twentieth century, after studying clay tablets. It also has become evident that the legendary accuracy of Babylonian observations took second place to mathematical theory in Babylonian astronomy.

A few tablets, in some instances mere fragments of tablets, reveal that the Babylonians recorded the positions of celestial objects and devised an arithmetical scheme to compute future positions. The tablets probably are from around the time of Hammurabi, in the eighteenth century before Christ.

Omens, such as the disappearances and reappearances of the planet Venus, also were recorded over several centuries, until about 1000 B.C. Systematic observational reports were made by the Assyrians around 700 B.C., and eclipse records have been found beginning in 747 B.C.

Systematic observations led to the development of a systematic mathematical theory, probably a little after 500 B.C. The Babylonians recognized that the moon completes 235 trips around the earth in the same time that the sun, viewed from the earth, seems to complete 19 cycles around the earth. This is a period of 19 years. The basic Babylonian solar year consisted of 12 lunar months, which in 19 years would make a total of 228 lunar months. To bring the total to 235, in 7 of the years an additional lunar month was inserted, or "intercalated".

The rule of 7 intercalations in 19 years remained the basis of lunar calendars through the Middle Ages. The concept of a number of intervals of one kind equaling a different number of intervals of another kind led to a more general mathematical astronomy.

In the Seleucid period, of the last three centuries B.C., the Babylonians developed the use of arithmetic progressions to describe the motions of celestial bodies. Over 10,000 tablets from apparently a single ancient archive reached the British Museum in the 1880s. Among the treasures were over a thousand texts and fragments of texts relating to astronomy. Of these, there are about 250 tablets of celestial positions and some 70 fragments of procedure texts. The fragments are too incomplete to reveal fully the arithmetic procedures employed by the Babylonians, but the procedures can be inferred or reconstructed from studies of tablets listing positions calculated by the procedures.

A table for the years 133/132 B.C. illustrates the use of mathematical progressions in calculating the position of the sun.

The Babylonians used a sexagesimal number system, based on the number 60. Positions were reported, for example, as 28, 55, 57, 58. Each succeeding unit represents so many 60ths of the preceding unit. The sexigesimal system undoubtedly is more cumbersome than our modern decimal system. But before we become too self-congratulatory, we should check the time.

The first column lists the months, from I through XII.

Numbers on a Babylonian Clay Tablet

month	solar motion	solar position	zodiac
XII	28, 55, 57, 58	22, 8, 18, 16	♈
I	28, 37, 57, 58	20, 46, 16, 14	♉
II	28, 19, 57, 58	19, 6, 14, 12	♊
III	28, 19, 21, 22	17, 25, 35, 34	♋
IV	28, 37, 21, 22	16, 2, 56, 56	♌
V	28, 55, 21, 22	14, 58, 18, 18	♍
VI	29, 13, 21, 22	14, 11, 39, 40	♎
VII	29, 31, 21, 22	13, 43, 1, 2	♏
VIII	29, 49, 21, 22	13, 32, 22, 24	♐
IX	29, 56, 36, 38	13, 28, 59, 2	♑
X	29, 38, 36, 38	13, 7, 35, 40	♒
XI	29, 20, 36, 38	12, 28, 12, 18	♓
XII	29, 2, 36, 38	11, 30, 48, 56	♈

The second column lists how far the sun moves in each month. During month I, the sun moves 28 degrees, 37 minutes, 57 seconds, and 58/60ths of a second (line 2, column 2). Add this movement to the position of the sun at the end of the previous month (line 1, column 3), and the result is the position of the sun at the end of month I (line 2, column 3).

The third column gives the position of the sun. It is obtained by adding to the initial position the amount of motion during the month, listed in the second column. For example, adding the top line of the third column and the second line of the second column produces the second line in the third column - after subtracting 30 degrees and listing the position 30 degrees ahead in the next segment of the zodiac.

The fourth column contains the zodiacal sign for the position of the sun that month. Each house of the zodiac occupies 30 degrees, one twelfth of the 360 degrees of a complete circle.

The table shows a decrease of 18 minutes in the distance traveled by the sun from one month to the next for months I and II of this particular year. For months III through VIII there is an increase of 18 minutes in the distance traversed each month. In the last four months, the distance is 18 minutes less in each consecutive month.

Within each of the three groups of months, the last two sets of numbers in each line of the second column, the solar motion, are constant. This seems suspicious. Obviously the numbers are not actual observations, but have been calculated by adding or subtracting 18 minutes from each preceding line.

The decreasing, increasing, and then decreasing again sequence of numbers in the second column, of the sun's motion, is called a zigzag function, after its appearance on a graph. Babylonian astronomers also used a system in which the solar velocity was assumed to remain constant for several months; then the sun would proceed with a different constant speed for several more months; and then the motion would revert to the initial velocity and remain constant at that speed for several more months. Graphed, the motion looks like a series of steps up and down, and it is called a step function.

Zigzag and step are useful labels, but potentially misleading. The Babylonians appear not to have used graphs. Nor did they construct geometrical models of the motions of celestial bodies. Nor were they concerned about the causes of the motions, at least not in the tablets studied so far. Nor did they express any curiosity about the physical composition of the celestial bodies.

Some historians of science accept the high level of mathematical theory in late Babylonian astronomy as the first appearance of modern science. Other historians, however, dismiss Babylonian astronomy as little more than a set of mechanical procedures, with no more theoretical content than recipes in a cook book. These historians look to the Greeks for the birth of modern science.

Debates over when and where "modern science" began depend partially upon how modern science is defined. Different features may be chosen as defining characteristics. Otto Neugebauer, a historian of ancient astronomy, rejects any emphasis on the underlying motives of science. He concludes that Babylonian astronomical texts of the Seleucid period are scientific because everything has been eliminated from the astronomy except observations and the mathematical consequences of an initial hypothesis about the fundamental character of the movements.

If the purpose is merely to hand out posthumous awards, there is little to gain from arguments over which ancient people first achieved which features of modern science. The question is useful, however, for stimulating comparisons of different features of science in different civilizations.

Ancient science is worthy of study for its own sake and as an important part of ancient civilization. Also, specific fragments of scientific knowledge are valuable as indicators of cultural contacts between civilizations.

While Babylonian astronomy in its broader sense of goals and methods did not become an accepted part of Western civilization, some aspects of Greek astronomy are traceable back to Babylonian origins. The historian of astronomy Asger Aaboe has found an instance in which Babylonian numerical values appear in the work of the late Greek astronomer Ptolemy.

Describing the work of his predecessor Hipparchus, Ptolemy wrote in about 150 A.D. that there are 4,267 synodic months in 126,007 days and 1 hour, and that the mean synodic month is therefore 29, 31, 50, 8, 20 days, that is, 29 plus 31/60 plus 50/3600 etc. days. (The mean synodic month is the period of revolution of the moon with respect to the sun. A sidereal month is measured with respect to the stars.)

However, dividing 126,007 days and 1 hour by 4267 produces not Ptolemy's value for the mean synodic month, but the number 29, 31, 50, 8, 9. Try it. If nothing else, the attempt produces an increased appreciation of the convenience of the decimal system!

It so happens that the Babylonian value of the mean synodic month is 29, 31, 50, 8, 20. Obviously there was an exchange of information between Babylonian mathematical astronomy and Greek astronomy. The exchange could have been part of a more general pattern of mutual influences between Hellenistic and Oriental civilizations.

Borrowing and competition among Babylonian and Greek astronomers may have stimulated developments in astronomy. Scientific exchanges may have contributed to the disintegration of old cultural traditions and opened the way for the creation of the new Roman civilization. New traditions established in astronomy may parallel new traditions in other areas of our cultural heritage. It remains for future historical studies to test these possibilities.

Bibliography

The best single book on ancient astronomy and mathematics is Otto Neugebauer's *Exact Sciences in Antiquity, second edition.* The table of numbers on a Babylonian clay tablet is reproduced from this book, with the permission of Dover Publications, Inc.

See also Dr. Neugebauer's article "The History of Ancient Astronomy: Problems and Methods" in the *Journal of Near Eastern Studies, volume 4* (1945), pages 1-38, reprinted with a new section on Jewish astronomy and an amplified bibliography in the *Publications of the Astronomical Society of the Pacific, volume 58* (1946), pages 17-43 and 104-142.

On Ptolemy's debt to Babylonian astronomy, see Asger Aaboe's article "On the Babylonian Origins of some Hipparchian Parameters" in the journal *Centaurus, volume 4* (1955), pages 122-125.

Edward Chiera describes Mesopotamian civilization and its discovery in his book *They Wrote on Clay.*

The film *Ancient Mesopotamia, revised edition* (10 minutes) was made on location and shows the legacies of that civilization, including accomplishments in astronomy. *Man and the Rise of Civilization* (19 minutes) discusses the agricultural revolution and the importance of abundant water and fertile soil.

A set of 21 slides on Babylonia and Babylonian science is available from the History of Science Collections, University of Oklahoma Libraries.

CHAPTER THREE

Matter and Number

The earliest human societies were composed of hunters and food gatherers. Old Stone Age or paleolithic man may have existed as much as half a million years ago. Around 10,000 B.C., agriculture and the domestication of animals began in a few regions, bringing in the New Stone Age or neolithic period. Neolithic culture began in the East, and spread to Greece during the fourth millennium B.C.

Indo-Europeans invaded Greece from the north around 2,000 B.C. and developed the Mycenaean civilization. At the same time, Minoan civilization was flourishing on the island of Crete. The Minoans were active traders, and they contributed to a diffusion of ideas from Egyptian and Asian civilizations to Greece.

Greek civilization and culture might well have remained strongly tied to the East. Minoan civilization, however, collapsed about 1400 B.C., probably because of Greek raids. The Mycenaean civilization soon was cut off from its Near-Eastern links by unrest in the Near East and by the Dorian invasion of Greece. Mycenaean kingdoms, trade, technology, and culture were swept away. Greece entered its own Dark Ages, beginning about 1200 B.C. and lasting until approximately 800 B.C.

When the level of civilization began to rise again, it was without the Oriental influences of the earlier period. The political base in Greece broadened and the city-state was developed as a form of government. New Greek cities were established throughout the Aegean region, including the coast of Asia Minor and around the Mediterranean. New trading opportunities were created and prosperity increased.

Scientific developments first appear in the new Greek world around 600 B.C. Distinctive approaches to the understanding of nature were formulated in the period of pre-Socratic science, before Socrates, from about 600 to 400 B.C. Science changed significantly but not completely at the end of the period, largely due to Socrates' influence.

Any discussion of pre-Socratic philosophy might well carry a warning notice that the conclusions drawn rest upon an uncertain foundation. Indeed, not a single original writing from the period has survived.

Knowledge of pre-Socratic science comes from three sources, all written well after the event:

(1) fragments, a few quotations from pre-Socratic works that have survived in books written after 400 B.C.;

(2) testimonia, comments in the writings, such as have survived, of Plato, Aristotle, and Theophrastus on pre-Socratic science. These books were written shortly after the pre-Socratic period;

(3) the doxographical tradition, consisting of summaries of the works of Plato, Aristotle, and Theophrastus, and summaries of summaries. The primary source for the summarizers was the multi-volume history of early philosophy by Theophrastus.

None of the sources is above question.

The fragments probably are the most reliable, or the least unreliable. But there is no surviving original against which to check the fragmented quotations.

The testimonia have an additional uncertainty. Aristotle gives serious attention to the earlier philosophers, but his judgment and his corresponding analysis and description of earlier philosophy may well have been distorted by his own belief in the importance of the material nature of the world. Plato, in contrast, offers only casual remarks. Of the 16 or 18 books written by Theophrastus, but a single volume has survived.

The doxographical tradition consists primarily of summaries of Theophrastus' books and summaries of summaries. The summaries are especially important because almost all of the original is lost. Summaries, though, are notoriously unreliable. The doxographical tradition is aptly named; "doxy" means both opinion and prostitute. Knowledge of the early philosophers almost certainly has been corrupted in the process of condensation, simplification, interpretation, translation, and transmission.

It may verge upon licentious behavior to base historical studies on obviously unreliable evidence. There is, however, no better evidence available.

One major tradition within pre-Socratic science was that established and developed by the Ionians. Greek communities sprang up on both sides of the Aegean Sea, in Greece and in Asia Minor. The most prosperous new city-state was Miletus, in the region of Ionia, in Asia Minor.

Now but lonely ruins inland from the Turkish coast because the river and harbor silted up, visited by an occasional tourist, Miletus was, in its time, the richest city in the Greek world. Ionians from Greece conquered earlier inhabitants, kept their women, and multiplied. Wealth followed trade, and Miletus at its height had some eighty colonies. In this environment philosophy flourished.

According to Aristotle, writing some two and a half centuries after the event, Milesian philosophers thought that matter or principles in the form of matter were the principle of all things.

Excerpt from Aristotle's *Metaphysics*

Most of the first philosophers thought that principles in the form of matter were the only principles of all things: for the original source of all existing things, that from which a thing first comes-into-being and into which it is finally destroyed, the substance persisting but changing in its qualities, this they declare is the element and first principle of existing things, and for this reason they consider that there is no absolute coming-to-be or passing away, on the ground that such a nature is always preserved . . . for there must be some natural substance, either one or more than one, from which the other things come-into-being, while it is preserved. Over the number, however, and the form of this kind of principle they do not all agree; but Thales, the founder of this type of philosophy, says that it is water (and therefore declared that the earth is on water), perhaps taking this supposition from seeing the nurture of all things to be moist, and the warm itself coming-to-be from this and living by this (that from which they come-to-be being the principle of all things) - taking the supposition both from this and from the seeds of all things having a moist nature, water being the natural principle of moist things.

The objective of Milesian science or philosophy (the two were not separate disciplines at this time) seems to have been to search for a basic substance or substances which persist throughout all changes.

Thales, the first of the Milesian philosophers mentioned by later writers, was reputed to be the wisest of the seven wise men or sages of Greece. Asked what was difficult, he answered "to know thyself". Asked what was easy, he answered "to give advice".

He was the first mathematician to demonstrate that a circle is bisected by its diameter, that the angle of a semicircle is a right angle, and that angles at the base of an isosceles triangle are equal.

Thales is said to have predicted a solar eclipse that occurred during the sixth year of a war and brought the war to an end. Either the warring parties took the eclipse as a sign to cease fighting, or they were eager for any reason to cease and found the eclipse a convenient excuse.

Prediction of the solar eclipse of 584 B.C. may be more myth than historic truth, however. From a study of the periodic recurrence of solar eclipses, Thales could have predicted an eclipse before the end of 583 B.C., and then taken credit for a different eclipse occurring slightly earlier.

Another legend involving Thales has him providing a practical justification for the study of philosophy. Reproached for his poverty, Thales determined to demonstrate to the skeptics that there were practical uses for philosophy or science. He studied the heavens and decided that weather conditions would produce a large olive crop the following year. He proceeded to raise money during the winter and rent all the olive presses. The next summer growers were forced to pay Thales his price to have their olives pressed into oil.

Thus did Thales demonstrate that philosophers could be rich in conventional monetary terms, if they wished. The philosopher's true wealth, however, is not measured in money. It is found in the pleasure derived from intellectual endeavor. In eschewing a myopic pursuit of wealth, Thales demonstrated again his wisdom.

Legends about Thales

Now Thales, as Herodotus and Douris and Democritus say, was the son of Examyes as father and Cleobuline as mother, from the descendants of Theleus, who are Phoenicians, nobles from the line of Cadmus and Agenor . . . and he [Agenor] was enrolled as a citizen in Miletus when he came with Neileos, when the latter was exiled from Phoenicia. But most people say that Thales was a true Milesian by descent, and of high family.

When he came to the Halys river, Croesus then, as I say, put his army across by the existing bridges; but, according to the common account of the Greeks, Thales the Milesian transferred the army for him. For it is said that Croesus was at a loss how his army should cross the river, since these bridges did not yet exist at this period; and that Thales, who was present in the army, made the river, which flowed on the left hand of the army, flow on the right hand also. He did so in this way: beginning upstream of the army he dug a deep channel, giving it a crescent shape, so that it should flow round the back of where the army was encamped, being diverted in this way from its old course by the channel, and passing the camp should flow into its old course once more. The result was that as soon as the river was divided it became fordable in both its parts.

Some think he was the first to study the heavenly bodies and to foretell eclipses of the sun and solstices, as Eudemus says in his history of astronomy; for which reason both Xenophanes and Herodotus express admiration; and both Heraclitus and Democritus bear witness for him.

Eudemus relates in the Astronomy *that Oenopides first discovered the obliquity of the Zodiac and the cycle of the Great Year, and Thales the eclipse of the sun and the variable period of its solstices.*

In the sixth year of the war, which they had carried on with equal fortunes, an engagement took place in which it turned out that when the battle was in progress the day suddenly became night. This alteration of the day Thales the Milesian foretold to the Ionians, setting as its limit this year in which the change actually occurred.

For when they reproached him because of his poverty, as though philosophy were no use, it is said, that, having observed through his study of the heavenly bodies that there would be a large olive crop, he raised a little capital while it was still winter, and paid deposits on all the olive presses in Miletus and Chios, hiring them cheaply because no one bid against him. When the appropriate time came there was a sudden rush of requests for the presses; he then hired them out on his own terms and so made a large profit, thus demonstrating that it is easy for philosophers to be rich, if they wish, but that it is not in this that they are interested.

. . . just as, Theodorus, a witty and attractive Thracian servant-girl is said to have mocked Thales for falling into a well while he was observing the stars and gazing upwards; declaring that he was eager to know the things in the sky, but that what was behind him and just by his feet escaped his notice.

The tale of the olive presses, true or not, demonstrates two different motives for the practice of science. The true philosopher pursues scientific inquiry for its own sake; others demand that science pay in practical results. Thales was not always practical. A witty slave girl mocked him for observing the stars rather than noticing what was at his feet, and consequently falling into a well.

For the basic principle of all things, Thales chose water. Aristotle guessed that Thales chose water because of its physiological importance. Or Thales might have been influenced by his interest in meteorology.

Thales began a tradition of searching for the basic matter or principle of things. Anaximander, probably fourteen years younger than Thales and described by Theophrastus as the successor and pupil of Thales, agreed with Thales that there was one basic principle of nature. But how could fire, destroyed by water, exist if water were the primary substance? Anaximander said that the first thing was not any specific substance, but something indefinite.

Accounts of Anaximander's Originative Substance

Version A	Version B	Version C
Of those who say that it is one, moving and infinite, Anaximander, son of Praxiades, a Milesian and the successor and pupil of Thales,	*Now Anaximander was the disciple of Thales. Anaximander, son of Praxiades, of Miletus:* . . .	*Anaximander, who was the companion of Thales,*
said that the principle and element of existing things was the aperion,	*he said that the principle and element of existing things was the aperion,*	*said that the aperion contained the whole cause of the coming-to-be and destruction*
being the first to introduce this name of the material principle	*being the first to use this name of the material principle*	
	In addition to this he said that motion was eternal, in which it results that the heavens come into being.	
He says that it is neither water nor any other of the so-called elements, but some other aperion nature,	*he said that the material principle of existing things was some nature coming under the heading of the aperion,*	

from which come into being all the heavens and the worlds in them.	*from which come into being the heavens and the world in them.*	*from which he says that the heavens are separated off, and in general all the worlds, being aperious,*
	This nature is eternal and unageing, and it also surrounds all the worlds.	
And the source of coming-to-be for existing things is that into which destruction, too, happens according to necessity;	*He talks of Time as though coming-to-be and existence and destruction were limited.*	*He declared that destruction and much earlier coming-to-be, happen from infinite ages, since they are all occurring in cycles.*
for they pay penalty and retribution to each other for their injustice according to the assessment of Time, as he describes it in these rather poetical terms.		

Next in the succession of pre-Socratic philosphers seeking the basic substance unchanged through all change was Anaximenes. He was twenty-four years younger than Anaximander and his associate in Miletus. For Anaximenes, air was the originative substance and basic form of matter, undergoing change by condensation and rarefaction.

Anaximenes on Nature

Anaximenes, son of Eurystratus, of Miletus, was a pupil of Anaximander; some say he was also a pupil of Parmenides. He said that the material principle was air and the infinite; and that the stars move, not under the earth, but round it. He used simple and economical Ionic speech. He was active, according to what Apollodorus says, around the time of the capture of Sardis, and died in the 63rd Olympiad.

Anaximenes and Diogenes make air, rather than water, the material principle above the other simple bodies.

Anaximenes, son of Eurystratus, of Miletus, a companion of Anaximander, also says that the underlying nature is one and infinite like him, but not undefined as Anaximander said but definite, for he indentifies it as air; and it differs in its substantial nature by rarity and density. Being made finer it becomes fire, being made thicker it becomes wind, then cloud, then (when thickened still more) water, then earth, then stones; and the rest come into being from these. He, too, makes motion eternal, and says that change, also, comes about through it.

Anaximander . . . said that infinite air was the principle, from which the things that are becoming, and that are, and that shall be, and gods and things divine, all come into being, and the rest from its products. The form of air is of this kind: whenever it is most equable it is invisible to sight, but is revealed by the cold and the hot and the damp and by movement. It is always in motion: for things that change do not change unless there be movement. Through becoming denser or finer it has different appearances; for when it is dissolved into what is finer it becomes fire, while winds, again, are air that is becoming condensed, and cloud is produced from air by felting. When it is condensed still more, water is produced; with a further degree of condensation earth is produced, and when condensed as far as possible, stones. The result is that the most influential components of generation are opposites, hot and cold.

For in the case of him [Anaximenes] alone did Theophrastus in the History *speak of rarefaction and condensation, but it is plain that the others, also, used rarity and density.*

. . . or, as Anaximenes thought of old, let us leave neither the cold nor the hot as belonging to substance, but as common dispositions of matter that supervene on changes; for he says that matter which is compressed and condensed is cold, while that which is fine and "relaxed" (using this very word) is hot. Therefore, he said, the dictum is not an unreasonable one, that man releases both warmth and cold from his mouth: for the breath is chilled by being compressed and condensed with the lips, but when the mouth is loosened the breath escapes and becomes warm through its rarity. This theory Aristotle claims to be due to the man's ignorance . . .

Afterwards, Anaximenes determined that air is a god, and that it comes into being, and is measureless and infinite and always in motion; as though either formless air could be a god . . . or mortality did not attend upon everything that has come into being.

Anaximenes (says that) the air (is god): one must understand, in the case of such descriptions, the powers which interpenetrate the elements or bodies.

He [Anaximander] left Anaximenes as his disciple and successor, who attributed all the causes of things to infinite air, and did not deny that there were gods, or pass them over in silence; yet he believed not that air was made by them, but that they arose from air.

Miletus was destroyed by the Persians in 494 B.C., but the tradition of searching for a basic principle of nature had spread to nearby cities. In Ephesus, just north of Miletus, Heraclitus around 500 B.C. argued that fire was the basic substance.

Heraclitus on the Nature of the Universe

Heraclitus, son of Bloson (or, according to some, of Herakon) of Ephesus. This man was at his prime in the 69th Olympiad. He grew up to be exceptionally haughty and supercilious, as is clear also from his book, in which he says: "Learning of many things does not teach intelligence; if so it would have taught Hesiod and Pythagoras, and again Xenophanes and Hecataeus." . . . Finally he became a misanthrope, withdrew from the world, and lived in the mountains feeding on grasses and plants. However, having fallen in this way into a dropsy he came down to town and asked the doctors in a riddle if they could make a drought out of rainy weather. When they did not understand he buried himself in a cow-stall, expecting that the dropsy would be evaporated off by the heat of the manure; but even so he failed to effect anything, and ended his life at the age of sixty.

Things taken together are whole and not whole, something which is being brought together and brought apart, which is in tune and out of tune; out of all things there comes a unity, and out of a unity all things.

God is day night, winter summer, war peace, satiety hunger [all the opposites, this is the meaning]; he undergoes alteration in the way that fire, when it is mixed with spices, is named according to the scent of each of them.

All things are an equal exchange for fire and fire for all things, as goods are for gold and gold for goods.

He does not reveal the nature of the surrounding; it contains, however, bowls turned with their hollow side towards us, in which the bright exhalations are collected and form flames, which are the heavenly bodies. Brightest and hottest is the flame of the sun . . . . And sun and moon are eclipsed when the bowls turn upwards; and the monthly phases of the moon occur as its bowl is gradually turned.

Further north of Miletus was Colophon. There Xenophanes, in the period 570 to 475 B.C., proposed that everything was composed of two elements, water and earth.

Xenophanes on the Nature of the Universe

Xenophanes, son of Dexios, or, according to Apollodorus, of Orthomenes, of Colophon . . . he, being expelled from his native land, passed his time in Zancle in Sicily and in Catana . . . . He wrote in epic meter, also elegiacs and iambics, against Hesiod and Homer, reproving them for what they said about the gods. But he himself also recited his own original poems. He is said to have held contrary opinions to Thales and Pythagoras, and to have rebuked Epimenides too. He had an extremely long life, as he himself somewhere says: "Already there are seven and sixty years tossing my thought up and down the land of Greece; and from my birth there were another twenty-five to add to these, if I know how to speak truly about these things." . . . And he was at his prime in the 60th Olympiad.

Of the Eleatic school Xenophanes the Colophonian is the pioneer, who Timaeus says lived in the time of Hieron, tyrant of Sicily, and the poet Epicharmus, while Apollodorus says that he was born in the 40th Olympiad and lasted until the times of Darius and Cyrus.

The sun comes into being each day from little pieces of fire that are collected, and the earth is infinite and enclosed neither by air nor by the heaven. There are innumerable suns and moons, and all things are made of earth.

He says that the sun and the stars come from clouds.

Xenophanes says that the sun is made of ignited clouds. Theophrastus in the Physical Philosophers *wrote that it is made of little pieces of fire collected together from the moist exhalation, and themselves collecting together the sun.*

Xenophanes said there are many suns and moons according to regions, sections, and zones of the earth, and that at a certain time the disc is banished into some section of the earth not inhabited by us, and so treading on nothing, as it were, produces the phenomenon of an eclipse. The same man says that the sun goes onwards ad infinitum, *but seems to move in a circle because of the distance.*

All things that come-to-be and grow are earth and water.

For we all come forth from earth and water.

Sea is the source of water, and source of wind; for neither would there be the force of wind blowing forth from inside clouds without the great ocean, nor river-streams nor the showery water from the upper air: but the great ocean is begetter of clouds and winds and rivers.

Xenophanes thinks that a mixture of the earth with the sea is going on, and that in time the earth is dissolved by the moist. He says that he has demonstrations of the following kind: shells are found inland and in the mountains, and in the quarries in Syracuse he says that an impression of a fish and of seaweed has been found, while an impression of a bay-leaf was found in Paros in the depth of the rock, and in Malta flat shapes of all marine objects. These, he says, were produced when everything was long ago covered with mud, and the impression was dried in the mud. All mankind is destroyed whenever the earth is carried down into the sea and becomes mud; then there is another beginning of coming-to-be, and this foundation happens for all the worlds.

Let these things be opined as resembling the truth . . . .

Map of the Greek World

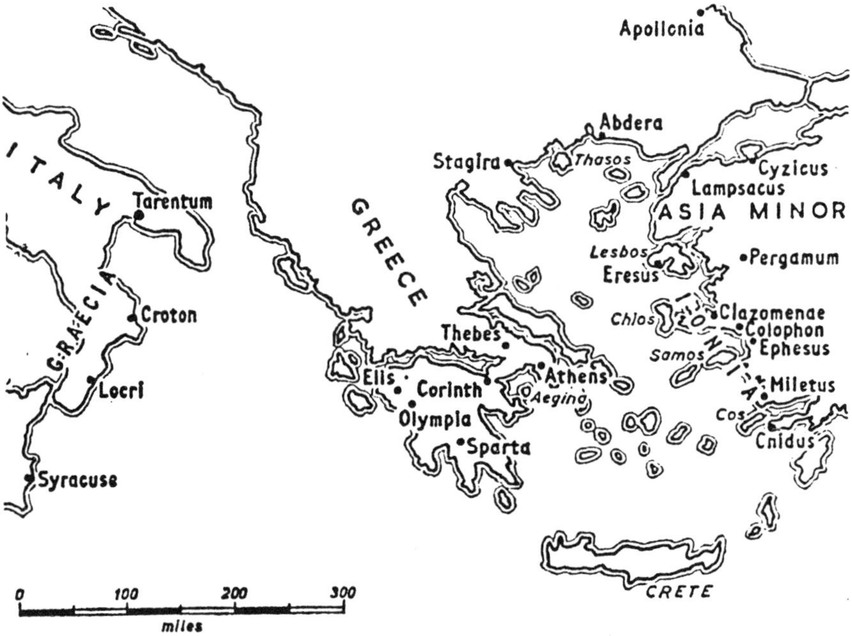

From Colophon, Xenophanes wandered to Sicily. There the tradition of seeking the basic substance pops up again. Xenophanes' doubling of the original single basic element was doubled again, by Empedocles. He lived from roughly 500 to 430 B.C. There were four basic elements for Empedocles: air, fire, earth, and water. Different mixtures of these elements produced different substances.

Empedocles on the Nature of the Universe

Empedocles, according to Hippobotus, was the son of Meton, himself son of Empedocles, and came from Acragas . . . . Eratosthenes, in his Olympic Victors, *says that the father of Meton won a victory in the seventy-first Olympiad, and he cites Aristotle as evidence. But Apollodorus the grammarian writes in his* Chronicles *that he was the son of Meton, and Glacus says that he came to Thurii very soon after its foundation . . . . According to Aristotle, and also Heraclides, he died at the age of sixty.*

Fools - for they have no far-reaching thoughts - who fancy that that which formerly was not can come into being or that anything can perish and be utterly destroyed. For coming into being from that which in no way is is inconceivable, and it is impossible and unheard-of that that which is should be destroyed. For it will ever be there wherever one may keep pushing it.

He makes the material elements four in number, fire, air, water and earth, all eternal, but changing in bulk and scarcity through mixture and separation; but his real first principles, which impart motion to these, are Love and Strife. The elements are continually subject to an alternate change, at one time mixed together by Love, at another separated by Strife; so that the first principles are, by his account, six in number.

Empedocles of Acragas . . . holds that the air that was separated off from the original mixture of the elements flowed around in a circle; and after the air, fire ran outwards and, having nowhere else to go, ran upwards under the solidified periphery around the air. There are, he says, two hemispheres revolving round the earth, one consisting entirely of fire, the other of a mixture of air with a little fire; this latter he supposes to be night. Their motion arises from the fact that the accumulation of fire in one region gives it preponderance there.

The sun is not in its nature fire, but rather a reflexion of fire like that which comes from water. The moon, he says, was composed of air that had been shut in by fire; this air was solidified, like hail. The moon gets its light from the sun.

Empedocles says that the stars are made of fire, composed of the fiery element which the air originally contained but squeezed out at the first separation.

All those who generate the heavens hold that it was for this reason that the earth came together to the center. They then seek a reason for its staying there; and some say, in the manner explained, that the reason is its size and flatness; others, like Empedocles, say that the motion of the heavens, moving about it at a higher speed, prevents movement of the earth, as the water in a cup, when the cup is given a circular motion, though it is often underneath the bronze, is for this same reason prevented from moving with the downward movement which is natural to it.

Theories about the supposed basic material nature of things constituted one major tradition of pre-Socratic thought. Generations of philosophers seemingly agreed that the basic material of nature was a subject worthy of their speculations, however unsubstantiated and lacking in practical consequence were those speculations.

A second major tradition within pre-Socratic philosophy centered about Pythagoras. He was born on the island of Samos, not far north of Miletus, probably between 585 and 565 B.C. He travelled in Asia, Greece, and Egypt in search of knowledge before settling at Croton in southern Italy in about 530 B.C. He migrated with his companions to Metapontium in about 495 B.C. There he died, probably between 495 and 475 B.C.

Pythagoras founded a secret religious fraternity. Its rule of secrecy has not helped historians intent on prying into the community's philosophy and the lives of the individual members. Also, the community was destroyed by enraged neighbors. Few Pythagoreans survived to tell their secrets, even had they wished to tell.

Legends about Pythagoras

Aristoxenus says that at the age of forty, seeing that the tyranny of Polycrates had grown more intense, . . . he eventually emigrated to Italy.

He emigrated to Croton in Italy and there, by legislating for the Italians, won renown together with his pupils. They numbered nearly 300, and they administered the affairs of state so well that the constitution was virtually an aristocracy.

Pythagoras departed to Metapontium, where he is said to have died.

Pythagoras, son of Mnesarchus, practiced scientific enquiry beyond all other men and, making a selection of these writings, claimed for his own a wisdom which was really dilettantism and malpractice.

After this his fame grew great, and he won many followers from the city itself (not only men but women also, one of whom, Theano, became very well known too) and many princes and chieftains from the barbarian territory around. What he said to his associates, nobody can say for certain; for silence with them was of no ordinary kind.

The strictness of their secrecy is astonishing; for in so many generations evidently nobody ever encountered any Pythagorean notes before the time of Philolaus; he first published those three notorious books, which Dion of Syracuse is said to have bought, at Plato's request, for 100 minae . . . .

Pythagoras wrote nothing, nor did Socrates nor Arcesilaus nor Carneades.

So Pythagoras turned geometrical philosophy into a form of liberal education by seeking its first principles in a higher realm of reality . . . .

Life, he said, is like a festival; just as some come to the festival to compete, some to ply their trade, but the best people come as spectators, so in life the slavish men go hunting for fame or gain, the philosophers for the truth.

The Pythagoreans, according to Aristoxenus, practiced the purification of the body by medicine, that of the soul by music.

The square on the hypotenuse of a right-angled triangle is equal to the sum of the squares on the sides enclosing the right angle. If we pay any attention to those who like to recount ancient history, we may find some of them referring this theorem to Pythagoras, and saying that he sacrificed an ox in honor of his discovery.

According to Aristotle, the Pythagoreans believed that number was the principle of all things.

Excerpt from Aristotle's *Metaphysics*

Contemporaneously with these philosophers, and before them, the Pythagoreans, as they are called, devoted themselves to mathematics; they were the first to advance this study, and having been brought up in it they thought its principles were the principles of all things.

Since of these principles numbers are by nature the first, and in numbers they seemed to see many resemblances to the things that exist and come into being - more than in fire and earth and water (such and such a modification of numbers being justice, another being soul and reason, another being opportunity - and similarly almost all other things being numerically expressible); since, again, they saw that the attributes and the ratios of the musical scales were expressible in numbers; since, then, all other things seemed in their whole nature to be modelled after numbers, and numbers seemed to be the first things in the whole of nature, they supposed the elements of numbers to be the elements of all things, and the whole heaven to be a musical scale and a number. And all the properties of numbers and scales which they could show to agree with the attributes and parts and the whole arrangement of the heavens, they collected and fitted into their scheme; and if there was a gap anywhere, they readily made additions so as to make their whole theory coherent. E.g. as the number 10 is thought to be perfect and to comprise the whole nature of numbers, they say that the bodies which move through the heavens are ten, but as the visible bodies are only nine, to meet this they invent a tenth - the "counter-earth". We have discussed these matters more exactly elsewhere.

Evidently, then, these thinkers also consider that number is the principle both as matter for things and as forming their modifications and their permanent states, and hold that the elements of number are the even and the odd, and of these the former is unlimited, and the latter limited; and the 1 proceeds from both of these (for it is both even and odd), and number from the 1; and the whole heaven, as has been said, is numbers.

Other members of this same school say there are ten principles, which they arrange in two columns of cognates - limit and unlimited, odd and even, one and plurality, right and left, male and female, resting and moving, straight and curved, light and darkness, good and bad, square and oblong. In this way Alcmaeon of Croton seems also to have conceived the matter, and either he got this view from them or they got it from him; . . . for he expressed himself similarly to them. For he says most human affairs go in pairs, meaning not definite contrarieties such as the Pythagoreans speak of, but any chance contrarieties, e.g. white and black, sweet and bitter, good and bad, great and small. He threw out indefinite suggestions about the other contrarieties, but the Pythagoreans declared both how many and which their contrarieties are.

From both these schools, then, we can learn this much, that the contraries are the principles of things; and how many these principles are and which they are, we can learn from one of the two schools. But how these principles can be brought together under the causes we have named has not been clearly and articulately stated by them; they seem, however, to range the elements under the head of matter; for out of these as immanent parts they say substance is composed and molded.

The selection by the Pythagoreans of intangible number as the principle of all matter may appear implausible to modern readers. There were, however, reasons in 500 B.C. to believe that number was important for an understanding of nature. Some of the reasons persist to today.

Musical harmony, for example, can be explained in terms of ratios of the lengths of strings or pipes. There are definite relationships between the lengths of vibrating strings and the pitch of notes emitted by strings. A ratio of 2 to 1 is an octave, 3 to 2 is a fifth, and 4 to 3 is a fourth.

Less plausible are attempts to understand qualities such as justice and goodness in terms of numerical ratios. Here the Pythagoreans seem far removed from the spirit of modern science. The search for numerical relationships among the phenomena of nature is, however, an important aspect of modern science. It can be traced to Pythagorean origins.

The Pythagorean emphasis on number changed history. Their ideas about the sun also influenced subsequent thought. They disagreed with the then-common belief that the earth lay in the center of the universe. Instead, they argued that the earth moved around a central fire. Pythagorean philosophy, including the displacement of the earth from the center of the universe and its movement about another body, may well have influenced Copernicus two millennia later, in the sixteenth century A.D. Copernicus accorded to the sun a central place in his planetary system. The Pythagoreans also were interested in the numerical harmonies of the motions of the planets. These were, in turn, related to musical harmonies. The planets supposedly moved in a manner such that the music their motions produced was harmonious.

Excerpt from Aristotle's *On the Heavens*

Most people say that the earth lies at the center of the universe. . . . but the Italian philosophers known as Pythagoreans take the contrary view. At the center, they say, is fire, and the earth is one of the stars, creating night and day by its circular motion about the center. They further construct another earth in opposition to ours to which they give the name counter-earth. In all this they are not seeking for theories and causes to account for observed facts, but rather forcing their observations and trying to accomodate them to certain theories and opinions of their own. But there are many others who would agree that it is wrong to give the earth the central position, looking for confirmation rather to theory than to the facts of observation. Their view is that the most precious place befits the most precious thing: but fire, they say, is more precious than earth, and the limit than the intermediate, and the circumference and the center are limits. Reasoning on this basis they take the view that it is not earth that lies at the center of the sphere, but rather fire. The Pythagoreans have a further reason. They hold that the most important part of the world, which is the center, should be most strictly guarded, and name it, or rather the fire which occupies that place, the "Guard-house of Zeus", as if the word "center" were quite unequivocal, and the center of this mathematical figure were always the same with that of the thing or the natural center. But it is better to conceive of the case of the whole heaven as analogous to that of animals, in which the center of the animal and that of the body are different.

From all this it is clear that the theory that the movement of the stars produces a harmony, i.e. that the sounds they make are concordant, in spite of the grace and originality with which it has been stated, is nevertheless untrue. Some thinkers suppose that the motion of bodies of that size must produce a noise, since on our earth the motion of bodies far inferior in size and in speed of movement has that effect. Also, when the sun and the moon, they say, and all the stars, so great in number and in size, are moving with so rapid a motion, how should they not produce a sound immensely great? Starting from this argument and from the observation that their speeds, as measured by their distances, are in the same ratios as musical concordances, they assert that the sound given forth by the circular movement of the stars is a harmony. Since, however, it appears unaccountable that we should not hear this music, they explain this by saying that the sound is in our ears from the very moment of birth and is thus indistinguishable from its contrary silence, since sound and silence are discriminated by musical contrast. What happens to men, then, is just what happens to coppersmiths, who are so accustomed to the noise of the smithy that it makes no difference to them.

However strange the doctrine of musical harmonies may appear today, it did furnish an essential stimulus for the scientific revolution of the sixteenth and seventeenth centuries. The astronomer Johannes Kepler spent much of his working life, which ended in 1630 A.D., in a search for the numerical ratios that philosophy convinced him existed in the arrangement of the planetary system. Kepler's famous three laws, now conveniently abstracted out of a morass of number mysticism and numerous less reputable "laws", are an essential part of modern science. Newton's law of gravity followed from Kepler's work.

In addition to Ionian materialism and Pythagorean mathematization, the thought of Parmenides holds interest for modern philosophy. Parmenides was born around 515 to 510 B.C. He raised disturbing questions about the unformulated assumptions of naive common sense, a characterization of much of the thought of earlier philosophers. Parmenides' criticism was a healthy check upon potential excesses of philosophical speculation. He made philosophers conscious of their assumptions.

Parmenides' criticism also stimulated the last great work of pre-Socratic philosophy, the atomic theory. The theory, though, had little effect upon subsequent thought. The appearance of an atomic theory in the scientific revolution of the sixteenth and seventeenth centuries A.D. was largely independent of the earlier version.

In pre-Socratic philosophy, nature was thought of as a comprehensive system ordered by law. Magical powers of gods were not invoked in explanations of nature. The expurgation of the supernatural has been seen as marking the transition from religion to philosophy.

Rational as pre-Socratic philosophy may have been, though, it was little more than unfounded speculation. Pre-Socratic philosophers lacked the necessary methodology and experimentation to advance beyond dogmatic statements about nature to a tested and verified account of nature.

There remained much scope for further developments in science, developments importantly influenced by pre-Socratic thought.

Bibliography

The starting point for a study of pre-Socratic philosophy is G. W. Kirk's and J. E. Raven's book, *The Presocratic Philosophers.* It contains the later Greek commentaries on the chief pre-Socratic philosophers and a discussion of the texts.

The readings are from the Kirk and Raven volume, with permission of the publisher and copyright holder, the Cambridge University Press.

On the rise of rational speculation, see F. M. Cornford's book *From Religion to Philosophy*.

A general evaluation of pre-Socratic science is presented by H. Cherniss in his article on "The Characteristics and Effects of Presocratic Philosophy", published in the *Journal of the History of Ideas, volume 12* (1951), pages 319-345.

Thales' eclipse prediction is discussed by Willy Hartner in "Eclipse Periods and Thales' Prediction of a Solar Eclipse - Historic Truth and Modern Myth", in the journal *Centaurus, volume 14* (1969), pages 60-71.

The film *Music of the Spheres* (52 minutes) in the *Ascent of Man* series deals with Pythagorean mathematical philosophy in its original setting, as well as its reappearances in Islamic and Renaissance times.

Early Aegean civilizations at Crete, Mycenae, and Troy, the subsequent Dark Ages, the Classical period, and the Macedonian conquests of Alexander the Great are all pictured in *The Greeks* (29 minutes).

The glory of the Greek civilization of the fifth century before Christ is shown in *Ancient World: Greece, Part II* (29 minutes).

On the constellations and their mythological origins, see *Constellations, Guides to the Night Sky* (11 minutes).

CHAPTER FOUR

Plato and Mathematical Astronomy

Pre-Socratic science set as its goal the discovery of the principles of nature, whether in the basic matter of the Ionians or in the mathematical ratios of the Pythagoreans. Pre-Socratic science of the sixth and fifth centuries B.C. was followed by a surge of new ideas in new directions. Hellenic science of the fourth century B.C. formulated new approaches to science. The Hellenic was followed by the Hellenistic, lasting from roughly 300 B.C. to 150 A.D. Hellenistic scientists developed in considerable detail the suggestions of their Hellenic predecessors.

During the Hellenic period, from 400 to 300 B.C., Plato and Aristotle set major directions for science. Their asperations and achievements were to dominate Western science for nearly two thousand years. The problems and procedures set by Plato for science, especially for astronomy, are the subject of this chapter.

The philosophy of an individual may be studied using the assumption or working hypothesis that philosophy is a response to the environment. Intellectual historians prefer this approach. Philosophers, in contrast, typically concentrate upon the ideas themselves, not biographical details or societal context.

Plato's philosophy might be viewed as a reaction to the temporary moral values of his age, which left Plato highly dissatisfied and, consequently, searching for a new philosophy. The change in Greek science instituted by Plato during the Hellenic period was but part of a larger change in Greek society. Plato's science followed from wider changes in civilization.

In 479 B.C., a year after the Persians under Xeres I had captured and burned Athens, thirty-one Greek city-states defeated the Persians in decisive land and sea battles. The victory brought to a successful resolution some twenty years of struggle to stop the westward expansion of the great Persian Empire. This was the beginning in Greece of a Golden Age.

Greece became increasingly dominated by Athens. The city's population rose to perhaps 400,000. Tribute poured in from other city-states, furnishing liberal support for Athenian writers and artists.

The subject city-states shared in the general prosperity, but resentment at the loss of political independence grew. Led by Sparta, several Greek city-states revolted against Athenian rule, setting off the Peloponnesian wars of 431-404 B.C.

Initially, Athens had no great difficulty maintaining dominance. The fortunes of battle shifted, however, after an unsuccessful attack by Athens on Syracuse, a city in Sicily, in 413 B.C. Athens was forced to surrender to Sparta in 404 B.C.

The next century was for Athens and other Greek city-states a period of decline. In contrast, the time of Athenian control truly was a Golden Age.

In the turmoil enveloping Greece, it is not surprising that a search for a new and more useful philosophy should occur. Ionian philosophy appeared dogmatic, not subject to experimental proof. A more serious perceived flaw in Ionian philosophy was its lack of relevance for the new age. The search for a material substance of nature was useless. The new condition of society demanded an attempt to rediscover the spiritual world of man's soul and the meaning of life.

Reaction against the materialistic trends of physical science was led by Socrates. His program of thought is summed up in the phrase "know thyself".

Socrates, who lived from 469 to 399 B.C., was trained as a stone cutter in his father's shop in Athens. He preferred, however, to spend his time arguing in the marketplace.

There he encouraged the youth of Athens to question every moral precept handed down to them. Elder citizens believed, with justification, that Socrates was demoralizing their children. Socrates' critical questioning also extended to the government and its acts. In addition to corrupting the youth of Athens, Socrates soon was accused of impiety.

The charge may have been intended to frighten him into fleeing Athens. But Socrates stayed and forced the issue. He welcomed his trial as a forum for his ideas.

A trial then consisted of two parts. In the first part, the guilt or innocence of the defendant was established. If guilty, the punishment was determined in a second part.

The jury found Socrates guilty. The prosecution proposed the punishment of death. Socrates proposed board and lodging at public expense because his actions had been for the public benefit.

The jury chose between the two proposed punishments. Socrates drank the fatal cup of hemlock.

Socrates' death and the related political conditions in Athens directly influenced Plato. He was Socrates' pupil and a close friend.

Born in 427 B.C., Plato was of an age to enter public life about the time of the defeat of Athens by Sparta in 404 B.C. Also, Plato's uncle was one of the thirty tyrants who ruled Athens immediatley after the defeat.

But the actions of the tyrants disgusted Plato. They quelled criticism by intimidation and opposition by assassination. They met treasury deficits by the arbitrary execution of wealthy individuals for treason, followed by confiscation of the alleged traitors' properties. Sensitive Plato chose not to join his uncle's government.

A year later, the democratic faction having driven out the tyrants, Plato again considered entering politics. Then the democracy executed Socrates.

Plato set aside political ambition and began his search for unchanging standards to hold against the shifting judgments of men. Subsequent experiences confirmed him in this course.

In 388 B.C. the dictator of Syracuse, Dionysius I, asked Plato if he didn't think that Dionysius was a happy man. According to legend, Plato answered that he thought no one who was not mad would become a tyrant. Enraged, Dionysius ordered Plato sold into slavery. A friend arrived just in time with ransom money to rescue Plato.

When Dionysius I died, an uncle of the young Dionysius II persuaded Plato to tutor the new tyrant. The uncle was soon banished, and the young dictator became impatient with the amount of effort demanded by his tutor. Plato had set out upon the ambitious task to educate Dionysius to become a philosopher-king. For some time Plato was not allowed to leave Syracuse. Later, he could leave only on the promise that he would return.

Plato tried to mediate the quarrel between Dionysius and his uncle. He was caught in the middle when negotiations broke down and barely escaped with his life.

Plato's philosophy and its implications for the study of astronomy are particularly understandable as a response to the time of troubles in which Plato found himself. Reacting to the temporary moral values of the time, Plato searched for unchanging standards. The changing, visible world was without permanent values. So Plato turned to the world of ideas. There he could hope to find the real and unchanging standards absent in the world of experience.

Plato's insistence that the only real world exists in the mind is implausible, at first glance. But the concept is plausibly illustrated with a simple example. Think of a circle and draw a circle. Which is real? The circle drawn on paper is not a real circle, no matter how skilled the draftsman. The drawn circle is an imperfect representation in the visible world of experience of a perfect circle. The perfect circle exists only in the mind, only in the world of thought.

Plato argued that reality exists not in the visible world but rather in the world of ideas. The implications of this argument for the study of astronomy are presented in Plato's book the *Republic*. Plato took up the question of how society might be shaped to bring out the best in man. He asked what justice is and how it is to be attained. Plato concluded that a philosopher-king is needed to bring the ideal state into existence. How such a person might be educated was one of Plato's concerns. In this context, the nature of reality and the study of astronomy are introduced.

In a parable to illustrate the degrees to which man may be enlightened, Plato presented his famous Allegory of the Cave. He imagined men chained from childhood in a cave. They are so shackled that they must sit still, looking in only one direction. A fire behind them casts shadows of objects upon the cave wall in front of them. In the absence of any other experience, the prisoners accept the shadows as reality. Furthermore, the objects casting the shadows are not real men, but statues of men.

Plato explained that the prison of the cave corresponds to the part of the world revealed by the sense of sight. Escape from the cave corresponds to the use of intelligence to reach the real world of knowledge.

In the *Republic*, as in most of Plato's books, the subject is presented in the form of a dialogue. Plato distrusted the fixed, dead words of textbooks. He believed that learning could be achieved only through discussion and shared inquiry.

The Allegory of the Cave

Now then, compare our natural condition, as far as education and ignorance are concerned, to the following. Imagine a number of men living in a cave. There they have been confined from birth, with their legs and necks shackled. They sit still and look straight forward at the cave wall. Their chains make it impossible for them to turn their heads around. Further imagine a bright fire burning above and behind them, and a road passing between the fire and the prisoners, with a low wall built along it, like the screens which conjurors put up in front of their audience and above which they exhibit their wonders.

I have it.

Also imagine a number of persons walking behind the wall. They carry statues of men and images of other animals. The statues are of wood and stone and all kinds of materials. Together with various other articles, they overtop the wall. And, as you might expect, let some of the passers-by be talking, and others silent.

You are describing a strange scene, and strange prisoners.

They resemble us. Let me ask you, in the first place, whether persons so confined could have seen anything of themselves or of each other beyond the shadows cast by the fire on the wall?

Certainly not, if you suppose them to have been imprisoned all their lives with their heads unmoved.

And is not their knowledge of the things carried past them equally limited?

Unquestionably so.

And if they were able to talk with one another, do you think they would give names to the shadows they see cast on the cave wall?

Doubtless they would.

And if the cave returned an echo from the wall in front of the prisoners whenever one of the passers-by spoke, let me ask you, what could the prisoners attribute the voice to, if not to the passing shadow?

Unquestionably they would refer it to that.

Then surely such persons would believe the shadows of manufactured articles to be the only realities.

Without a doubt they would.

Now consider what would happen were the prisoners released from their fetters and their foolishness remedied in the following manner. Let us suppose that one of them has been released and forced suddenly to stand up and turn his neck around and walk with open eyes toward the light. And let us suppose that his actions are painful and that the dazzling splendor leaves him unable to see the shadows he formerly saw. What answer would you expect him to make if someone were to tell him that he was formerly watching foolish phantoms but now he is somewhat nearer reality, is turned toward things more real, sees more correctly? Above all, if he were asked what the several objects passing by were, should you not expect him to be puzzled and to regard his old visions as truer than the objects now forced upon his notice?

Yes, much truer.

And if he were further forced to stare at the light itself, would not his eyes be hurt? And would he not shrink, and turn back to the shadows which he could see distinctly? And would he not believe the shadows more clear and more real than the real objects?

Just so.

And if he were dragged violently up the rough and steep ascent from the cave into the sunlight, would he not be vexed and indignant at such treatment? And on reaching the light would he not be blinded by its radiance and unable to see any of the things he was now told were real? And would he not be ready to go through anything rather than entertain these new opinions and live in this new way?

For my part, I am quite of that opinion.

Now consider what would happen if he were to descend again and resume his old seat. Would he not find it difficult to see anything in the gloom of the cave?

Certainly he would.

And if he were forced, before his eyes had adapted to the dim light, to compete with the prisoners in describing the shadows, would they not laugh at him and say that he had gone up only to come back with his eyesight destroyed and that it was not worth while even to attempt the ascent? And if anyone endeavored to set them free and drag them out of the cave to the light, would they not ill treat that person if they could get him in their power?

Yes, they would. They would kill him.

Now apply this imaginary case to our former statements. Compare the region which the eye reveals to the cave. Compare the light of the fire to the light of the sun. Compare the upward ascent to the contemplation of the upper world. Now you understand the mounting of the soul to the intellectual region. In the world of knowledge the essential Form of Good is the limit for our inquiries and can barely be perceived. But, when perceived, we cannot but conclude that it is in every case the source of all that is bright and beautiful - in the visible world giving birth to light and in the intellectual world dispensing truth and reason. Whosoever would act wisely, either in private or in public, must set this Form of the Good before his eyes.

In the allegory of the cave, knowledge is presented as the power that will set men free, free from the shackles of ignorance and free to live enlightened lives in the sunlight of understanding.

The next issue in the *Republic* is how to bring man to a state of enlightenment, how to convert the soul. Education is the answer. Education will enable the soul to look in the right direction and learn the truth.

The forms of study leading the soul away from the world of change and to the world of reality are the sciences. They deal with certain knowledge and immutable truths. Arithmetic, geometry, solid geometry, astronomy, and harmonics quicken the mind and help a pupil gain understanding of the essential form of goodness - so Plato believed.

Studying these subjects for ten years would release the prisoner from the cave and enable him to look at shadows and reflections. Then there must follow an additional five years of study, devoted to dialectics.

Plato, enjoying the benefits of education and committed to education, entertained no doubt about the enthusiasm of students. Before freed from the cave of ignorance, however, students know no other nor better world, nor reason for the laborious effort required to obtain an education. Some juveniles, no doubt, might think they would prefer the prison of a cave and ignorance to the prison of a school and all the effort inherent in the process of becoming educated. Probably that's how the young Dionysius felt when he became impatient with Plato's program of education.

In discussing astronomy, Plato first mentioned its utilitarian benefits. Agriculture, navigation, and war are alluded to. It is not for these purposes, however, that astronomy is to be esteemed. The true utility of the regimen of study prescribed in the *Republic* is the saving of the soul.

An obvious approach to astronomy is to observe the motions of the objects in the heavens. But only a discipline dealing with unseen reality will lead the mind upward. The true motions are not to be seen with the eye. It is not by looking at the heavens that one can become truly acquainted with astronomy.

The Study of Astronomy

Shall we study astronomy? Shall we make it a study in our curriculum? Or do you disapprove?

Yes, I approve. To know the seasons, the months, and the years is important for the farmer and the sailor, and also for the general.

You amuse me by your evident fear that the common man might think that you insist upon useless studies. It is not easy to appreciate the true worth of astronomy. It is very difficult to believe that in the midst of these studies an organ of our souls is being purged from blindness and quickened from deadness. This organ is more important than a thousand eyes. Only by it can truth be seen. Consequently, those who share our understanding will endorse without qualification the studies you prescribe. But those with no inkling at all of this doctrine will think the studies valueless because they see no considerable advantage to be gained from them beyond their practical applications. Therefore, consider at once which audience you try to appeal to. Or if you are carrying on the discussion chiefly for your own benefit without any attempt to persuade others, you surely will not grudge another man any advantage which he may derive from conversation.

I prefer the latter course.

Then take a step backward. We were wrong a moment ago in what we took as the science next in order to geometry.

What did we take?

We proceeded to solids in revolution before considering solids themselves.

True, but these subjects have not yet been explored.

They have not, for two reasons. In the first place, they are difficult problems but feebly investigated because no city holds them in esteem. In the second place, those who do investigate them are in need of a superintendent. To find such a person is a hard task. And supposing one were found. As matters stand, those who are inquisitive about the subject have too much pride to allow them to listen to suggestions. But if a city were to pay honor to the study and constitute itself superintendent, students then would pay attention. The real nature of the subject, continuously and vigorously investigated, would be brought to light. For even now, slighted and curtailed as it is not only by the many but also by professed inquirers who can give no account of the extent of its usefulness, it nevertheless makes progress in spite of all these obstacles, by its inherent elegance.

There certainly is a peculiar fascination about it. You proceeded to place astronomy next to geometry, though afterward you drew back.

Yes. I said the more I haste to travel over the ground, the worse I speed. The investigation of three dimensions succeeds to plane geometry. But because it is studied absurdly I passed it over and spoke of astronomy, which implies motion of solid bodies, as the next step after geometry.

You are right.

Then let us assign the fourth place in our studies to astronomy.

It is a reasonable idea. And to return to the rebuke which you gave me a little while ago for my vulgar commendation of astronomy, I can now praise the plan on which you pursue it. For I suppose it is clear to everyone that astronomy, at all events, compels the soul to look upward and draws it from the things of this world to the other.

It may be clear to everyone else, but it is not clear to me.

What is your opinion?

It seems to me that astronomy, as now handled by those who study philosophy, positively makes the soul look downward.

How so?

I think you have betrayed no want of intrepidity in the conception you have formed of the true nature of that learning which deals with the things above. For probably if a person were to throw his head back and learn something from the contemplation of a carved ceiling, you would suppose him to be contemplating it not with his eyes but with his reason. Now perhaps your notion is right; perhaps mine is foolish. But I cannot conceive that any science makes the soul look upward unless it has to do with the real. And the real is invisible. It matters not whether a person stares at the ground or at the heavens. As long as he is trying to study any sensible object, he cannot be said to have learned anything, because no objects of sense admit of scientific treatment. And I maintain that his soul is looking downward, not upward, though he may be lying on his back staring upward.

I am rightly punished, for I deserved your rebuke. But what did you mean by saying that astronomy ought to be studied on a system very different from the present one if it is to be studied profitably for the purposes that we have in view?

I will tell you. Since this decorated sky is still a part of the visible world, we are bound to regard it, though the most beautiful and perfect of visible things, as far inferior to those true revolutions which real velocity and real slowness, existing in true number and in all true forms, accomplish relatively to each other, carrying with them all that they contain, which are truly discerned by reason and thought, but not by sight. Or do you think differently?

No, indeed.

Therefore we must use observations of this decorated sky as a pattern or plan to forward the study which aims at those higher objects, just as we might use the drawings of a draftsman as a model or pattern of the true and real but not reality itself. For, I imagine, a person acquainted with geometry, on seeing such diagrams, would think them most beautifully finished, but would regard it as ridiculous to study them seriously in the hope of detecting the truths of equality or duplicity or any other ratio.

No doubt it would be ridiculous.

And do you not think that the genuine astronomer will view with the same feelings the motions of the stars? That is to say, will he not regard the heaven itself and the bodies which it contains as framed by the heavenly architect with the utmost beauty of which such works are susceptible? But as to the proportion which the day bears to the night, both to the month, the month to the year, and the other stars to the sun and moon and to one another - will not the genuine astronomer look with scorn upon whomever believes such corporeal and visible objects to be changeless and exempt from all perturbations. And will he not believe it absurd to devote extraordinary effort to the attempt to observe and record the motions of the stars, sun, and moon?

Yes, I think so, now that I hear you suggest it.

Therefore we shall study astronomy with the help of problems, just as we study geometry. But we shall let the heavenly bodies alone, if it is our design to become really acquainted with astronomy and by that means to convert the natural intelligence of the soul from a useless into a useful possession.

Plato's disdain for observation, his near contempt for observation, is readily apparent in his comments on the study of astronomy and has not been generally perceived as beneficial to the progress of science. Plato's denial of the value of the senses in scientific studies has contributed to a generally negative view of him as a scientist. In any emphasis on the empirical part of science, in which facts are observed, Plato fares poorly.

The formation of theories is also an important aspect of science. In this context, Plato shines more favorably. If modern science is characterized as the discovery of mathematical relations between phenomena, Plato becomes a hero.

In his emphasis on mathematics, Plato stands close to the Pythagoreans. The two philosophies are often viewed as one.

In the *Republic* Plato tells how to study astronomy and how not to study it. Plato's thoughts about the actual structure of the astronomical universe are presented in the *Timaeus*, his dialogue on nature.

It is a difficult book. It seems more a beautiful story by a poet or mystic than a serious scientific work. Or, if it is a scientific work, it is an aberration.

Plato attempts to reveal the operation of reason in the world. He discusses both the unchanging and eternally existing forms or ideas and the actual world which is changing, the world of becoming.

Here the example of the idea of a circle and the imperfect drawing of a circle may be a helpful analogy.

Because the form and the actual are not identical, Plato's cosmology may appear at times to be inconsistent.

The *Timaeus* has also been criticized as more a collection of ideas of other philosophers than a work of original scholarship. To the extent that the *Timaeus* is a summary of prevailing opinion, it is a valuable historical document. To the extent that the *Timaeus* differs from modern scientific reasoning, it is a valuable view of a very different science practiced in ancient Greece.

The setting of the *Timaeus* is a banquet. On the previous day Socrates had discussed what kind of republic was best and what sort of men should compose such a republic. Now Timaeus, the most astronomical of the four men at the banquet, is asked to speak on the generation of the world.

He begins with the distinction between real being, in this case the plan of the universe apprehended by intelligence and reason, and the current state of things, never in any eternal time scale and perceived by irrational sense.

The object of scientific study is not a description of the world of the senses but the discovery of the plan of the Creator of the world. Confusion and disorder are bad; therefore the Divinity fabricated an orderly and hence beautiful universe. Anything moving with confusion and disorder was evil, and to be reduced to order.

Furthermore, since the world is perfect, there is only one world. The reasoning leading to the conclusion that only one universe exists may be unfamiliar to modern readers.

Also probably unfamiliar is the reasoning by which Timaeus concludes that the universe consists of fire, earth, water, and air, why the universe has a spherical shape, and why it has circular motion. But however unfamiliar and implausible now, this type of reasoning apparently was persuasive in Plato's time.

Timaeus further concludes that the Divinity generated the sun, moon, and planets to demonstrate the flow of time. Here the final cause serves as explanation of the phenomena.

The planets move in circular orbits, with the circles apparently varying in size. The orbits also revolve with a spiral motion that produces the wandering motion of the planets. (There are brief periods of retrograde motion, during which a planet seemingly stops its forward movement to move backward.)

Plato's suggestion here is vague, and but slightly suggestive of the epicycle theory soon to be developed. In the epicycle theory, a large circle (the deferent) carries a small circle (the epicycle). The epicycle revolves about a point on the circumference of the large circle. The planet is attached to the circumference of the small circle. This combination of circles varies the distance of a planet from the center of the large circle and produces the observed temporary retrograde motion.

Excerpts from the *Timaeus*

Critias: Then I will submit to you the plan we have arranged for your entertainment, Socrates. We decided that Timaeus shall speak first. He knows more of astronomy than the rest of us and has made knowledge of the nature of the universe his chief object; he will begin with the birth of the world and end with the nature of man.

Socrates: I see that I am to receive a complete and splendid banquet of discourse in return for mine. So you, Timaeus, are to speak next, when you have invoked the gods as custom requires.

Timaeus: Be that as it may, we must go back to this question about the world: After which of the two models did its builder frame it - after that which is always in the same unchanging state, or after that which has come to be? Now if this world is good and its maker is good, clearly he looked to the eternal; on the contrary supposition (which cannot be spoken without blasphemy), to that which has come to be. Everyone, then, must see that he looked to the eternal; for the world is the best of things that have become, and he is the best of causes. Having come to be, then, in this way, the world has been fashioned on the model of that which is comprehensible by rational discourse and understanding and is always in the same state.

Now that which comes to be must be bodily, and so visible and tangible; and nothing can be visible without fire, or tangible without something solid, and nothing is solid without earth. Hence the god, when he began to put together the body of the universe, set about making it of fire and earth. But two things alone cannot be satisfactorily united without a third; for there must be some bond between them drawing them together. And of all bonds the best is that which makes itself and the terms it connects a unity in the fullest sense; and it is of the nature of a continued geometrical proportion to effect this most perfectly. For whenever, of three numbers, the middle one between any two that are either cubes or squares is such that, as the first is to it, so is it to the last, and conversely as the last is to the middle, so is the middle to the first, then since the middle becomes first and last, and again the last and first become middle, in that way all will necessarily come to play the same part toward one another, and by so doing they will all make a unity.

Now if it had been required that the body of the universe should be a plane surface with no depth, a single mean would have been enough to connect its companions and itself; but in fact the world was to be solid in form, and solids are always conjoined, not by one mean, but by two. Accordingly the god set water and air between fire and earth, and made them, so far as was possible, proportional to one another, so that as fire is to air, so is air to water, and as air is to water, so is water to earth, and thus he bound together the frame of a world visible and tangible.

For these reasons and from such constituents, four in number, the body of the universe was brought into being, coming into concord by means of proportion, and from these it acquired Amity, so that coming into unity with itself it became indissoluble by any other save him who bound it together.

And for shape he gave it that which is fitting and akin to its nature. For the living creature that was to embrace all living creatures within itself, the fitting shape would be the figure that comprehends in itself all the figures that are; accordingly, he turned its shape rounded and spherical, equidistant every way from center to extremity - a figure the most perfect and uniform of all; for he judged uniformity to be immeasurably better than its opposite. And all round on the outside he made it perfectly smooth, for several reasons. It had no need of eyes, for nothing visible was left outside; nor of hearing, for there was nothing outside to be heard. There was no surrounding air to require breathing, nor yet was it in need of any organ by which to receive food into itself or to discharge it again when drained of its juices. For nothing went out or came into it from anywhere, since there was nothing: it was designed to feed itself on its own waste and to act and be acted upon entirely by itself and within itself; because its framer thought that it would be better self-sufficient, rather than dependent upon anything else.

It had no need of hands to grasp with or to defend itself, nor yet of feet or anything that would serve to stand upon; so he saw no need to attach to it these limbs to no purpose. For he assigned to it the motion proper to its bodily form, namely, that one of the seven which above all belongs to reason and intelligence; accordingly, he caused it to turn about uniformly in the same place and within its own limits and made it revolve round and round; he took from it all the other six motions and gave it no part in their wanderings. And since for this revolution it needed no feet, he made it without feet or legs.

All this, then, was the plan of the god who is forever for the god who was sometime to be. According to this plan he made it smooth and uniform, everywhere equidistant from its center, a body whole and complete, with complete bodies for its parts. And in the center he set a soul and caused it to extend throughout the whole and further wrapped its body round with soul on the outside; and so he established one world alone, round and revolving in a circle, solitary but able by reason of its excellence to bear itself company, needing no other acquaintance or friend but sufficient to itself. On all these accounts the world which he brought into being was a blessed god.

And having made a unity of the three, again he divided this whole into as many parts as was fitting, each part being a blend of Sameness, Difference, and Existence.

And he began the division in this way. First he took one portion (1) from the whole, and next a portion (2) double of this; the third (3) half as much again as the second, and three times the first; the fourth (4) double of the second; the fifth (9) three times the third; the sixth (8) times the first; and the seventh (27) twenty-seven times the first.

Next, he went on to fill up both the double and the triple intervals, cutting off yet more parts from the original mixture and placing them between the terms, so that within each interval there were two means, the one (harmonic) exceeding the one extreme and being exceeded by the other by the same fraction of the extremes, the other (arithmetic) exceeding the one extreme by the same number whereby it was exceeded by the other.

These links gave rise to intervals of 3/2 and 4/3 and 9/8 within the original intervals.

And he went on to fill up all the intervals of 4/3 (i.e., fouths) with the interval 9/8 (the tone), leaving over in each a fraction. This remaining interval of the fraction had its terms in the numerical proportion of 256 to 243 (semitone).

By this time the mixture from which he was cutting off these portions was all used up.

This whole fabric, then, he split lengthwise into two halves; and making the two cross one another at their centers in the form of the letter X, he bent each round into a circle and joined it up, making each meet itself and the other at a point opposite to that where they had been brought into contact.

He then comprehended them in the motion that is carried round uniformly in the same place, and made the one the outer, the other the inner circle. The outer movement he named the movement of the Same; the inner, the movement of the Different. The movement of the Same he caused to revolve to the right by way of the side; the movement of the Different to the left by way of the diagonal.

And he gave the supremacy to the revolution of the Same and uniform; for he left that single and undivided; but the inner revolution he split in six places into seven unequal circles, severally corresponding with the double and triple intervals, of each of which there were three. And he appointed that the circles should move in opposite senses to one another; while in speed three would be similar, but the other four should differ in speed from one another and from the three, though moving according to ratio.

When the whole fabric of the soul had been finished to its maker's mind, he next began to fashion within the soul all that is bodily, and brought the two together, fitting them center to center. And the soul, being everywhere inwoven from the center to the outermost heaven and enveloping the heaven all round on the outside, revolving within its own limit, made a divine beginning of ceaseless and intelligent life for all time.

In virtue, then, of this plan and intent of the god for the birth of Time, in order that Time might be brought into being, Sun and Moon and five other stars - "wanderers", as they are called - were made to define and preserve the numbers of Time. Having made a body for each of them, the god set them in the circuits in which the revolution of the Different was moving - in seven circuits seven bodies: the Moon in the circle nearest the Earth; the Sun in the second above the Earth; the Morning Star (Venus) and the one called sacred to Hermes (Mercury) in circles revolving so as, in point of speed, to run their race with the Sun, but possessing the power contrary to his; whereby the Sun and the star of Hermes and the Morning Star alike overtake and are overtaken by one another. As for the remainder, where he enshrined them and for what reasons - if one should explain all these, the account, though only by the way, would be a heavier task than that for the sake of which it was given. Perhaps these things may be duly set forth later at our leisure.

To resume: when each one of the beings that were to join in producing Time had come into the motion suitable to it, and, as bodies bound together with living bonds, they had become living creatures and learned their appointed task, then they become to revolve by way of the motion of the Different, which was aslant, crossing the movement of the Same and subject to it: some moving in greater circles, some in lesser; those in the lesser circles moving faster, those in the greater more slowly.

So, by reason of the movement of the Same, those which revolve most quickly appeared to be overtaken by the slower, though really overtaking them. For the movement of the Same, which gives all their circles a spiral twist because they have two distinct forward motions in opposite senses, made the body which departs most slowly from itself - the swiftest of all movements - appear as keeping pace with it most closely.

And in order that there might be a conspicuous measure for the relative speed and slowness with which they moved in their eight revolutions, the god kindled a light in the second orbit from the Earth - what we now call the Sun - in order that he might fill the whole heaven with his shining and that all living things for whom it was meet might possess number, learning it from the revolution of the Same and uniform. Thus and for these reasons day and night came into being, the period of the single and most intelligent revolution.

The month comes to be when the Moon completes her own circle and overtakes the Sun; the year, when the Sun has gone round his own circle. The periods of the rest have not been observed by men, save for a few; and men have no names for them, nor do they measure one against another by numerical reckoning. They barely know that the wanderings of these others are time at all, bewildering as they are in number and of surprisingly intricate pattern. Nonetheless it is possible to grasp that the perfect number of time fulfills the perfect year at the moment when the relative speeds of all the eight revolutions have accomplished their courses together and reached their consummation, as measured by the circle of the Same and uniformly moving.

In this way, then, and for these ends were brought into being all those stars that have turnings on their journey through the Heaven; in order that this world may be as like as possible to the perfect and intelligible Living Creature, in respect of imitating its ever-enduring nature.

Now so far, up to the birth of Time, the world had been made in other respects in the likeness of its pattern; but it was still unlike in that it did not yet contain all living creatures brought into being within it. So he set about accomplishing this remainder of his work, making the copy after the nature of the model. He thought that this world must possess all the different forms that intelligence discerns contained in the Living Creature that truly is. And there are four: one, the heavenly race of gods; second, winged things whose path is in the air; third, all that dwells in the water; and fourth, all that goes on foot on the dry land.

The form of the divine kind he made for the most part of fire, that it might be most bright and fair to see; and after the likeness of the universe he gave them well-rounded shape, and set them in the intelligence of the supreme to keep company with it, distributing them all round the heaven, to be in very truth an adornment for it, embroidered over the whole. And he assigned to each two motions: one uniform in the same place, as each always thinks the same thoughts about the same things; the other a forward motion, as each is subjected to the revolution of the Same and uniform. But in respect to the other five motions he made each motionless and still, in order that each might be as perfect as possible.

For this reason came into being all the unwandering stars, living beings divine and everlasting, which abide forever revolving uniformly upon themselves; while those stars that having turnings and in that sense "wander" came to be in the manner already described.

And Earth he designed to be at once our nurse and, as she winds round the axis that stretches right through, the guardian and maker of night and day, first and most venerable of all the gods that are within the heaven.

To describe the evolutions in the dance of these same gods, their juxtapositions, the counterrevolutions of their circles relatively to one another, and their advances; to tell which of the gods come into line with one another at their conjunctions, and which in opposition, and in what order they pass in front of or behind one another, and at what periods of time they are severally hidden from our sight and again reappearing send to men who cannot calculate panic fears and signs of things to come - to describe all this without visible models of these same would be labor spent in vain. So this much shall suffice on this head, and here let our account of the nature of the visible and generated gods come to an end.

Plato frequently explained phenomena as the result of final causes. Whatever is generated is necessarily generated from some cause. The world was generated according to an idea in the mind of the artificer. The cause of the world coming into being is the idea. The world is generated to fulfill the idea.

The use of final causes as explanation for phenomena is called teleology. Pervasive in the *Timaeus* and throughout much of Greek science, teleology is at odds with modern scientific practice. Modern man seeks mechanical actions or forces as explanation of the phenomena he observes. The teleological aspect of Plato's science is yet another factor contributing to his current low esteem as a scientist.

Whatever valuation is placed upon Plato's ideas, there remains the historical consideration of the influence of his ideas on the ideas of others. Regardless of intrinsic merit or flaw in Plato's philosophy, what was its actual impact on history?

Plato's ideas were presented in books and were taught directly to pupils. Plato founded in Athens in about 380 B.C. the Academy as a place of research and writing. Surviving until 529 A.D., the Academy was the world's first university.

Eudoxus, the greatest genius in mathematics and astronomy of his time, attended Plato's lectures at the Academy. Upon his report of what Plato said a string of statements followed, each based on a previous statement.

Eudoxus' report is lost. However, it was summarized by Eudemus in his own *History of Astronomy*.

This work also is lost. It was commented upon, though, by Sosigenes, who lived in the time of Julius Caesar and advised Caesar on the calendar reform of 45 B.C.

Sosigenes' work is lost, too. It was used, though, by Simplicius of Athens in the sixth century A.D.

Here the string of lost but summarized works ends. Simplicius' work survived.

In his commentary on Aristotle's *On the Heavens*, Simplicius stated that Plato set for astronomers the principle that the heavenly bodies move in circular orbits with uniform speed. From this principle followed the problem of determining what combination of uniform circular motions could be devised to agree with the observed motions. The task for astronomers working in the Platonic tradition thus was to "save the appearances", to explain apparently irregular motions detected by the senses as a combination of uniform circular motions, the real motions or forms ascertained by reason and thought.

Simplicius' *Commentary* on *On the Heavens*

Plato lays down the principle that the heavenly bodies' motion is circular, uniform, and constantly regular. Thereupon he sets the mathematicians the following problem: what circular motions, uniform and perfectly regular, are to be admitted as hypotheses so that it might be possible to save the appearances presented by the planets?

The curious problem of astronomers is the following: first, they provide themselves with certain hypotheses: the ancients, the contemporaries of Eudoxus and Callippus, adopted the hypothesis of "counterturning spheres"; Aristotle, who in his Metaphysics *teaches the system of spheres, must be counted among them. The astronomers who followed proposed the hypothesis of eccentrics and epicycles. Starting from such hypotheses, astronomers then try to show that all the heavenly bodies have a circular and uniform motion, that the irregularities which become manifest when we observe these bodies - their now faster, now slower motion; their moving now forward, now backward; their latitude now southern, now northern; their various stops in one region of the sky; their at one time seemingly greater, and at another time seemingly smaller diameter - that all these things and all things analogous are but appearances and not realities.*

To save these irregularities, astronomers imagine that each star is moved by several motions at the same time - some assuming movements along eccentrics and epicycles, others appealing to spheres homocentric with the world (the so-called counterturning spheres). But just as the stops and retrograde motions of the planets are, appearances notwithstanding, not viewed as realities (they are no more real than the numerical additions and subtractions with which we meet in studying these motions), so an explanation which conforms to the facts does not imply that the hypotheses are real and exist. By reasoning about the nature of the heavenly movements, astronomers were able to show that these movements are free from all irregularity, that they are uniform, circular, and always in the same direction. But they have been unable to establish in what sense, exactly, the consequences entailed by these arrangements are merely fictive and not real at all. So they are satisfied to assert that it is possible, by means of circular and uniform movements, always in the same direction, to save the apparent movements of the wandering stars.

To save the appearances may appear to modern readers an arbitrary and absurd task. No doubt the observed irregular motions of the planets could be reproduced using, in clever combination, a near-infinite number of perfect circles of various size with perfect, unchanging velocities. But it would be a cumbersome contraption.

Furthermore, it would tell little of the cause of the motions. It would tell little unless one believed in the physical actuality in the world of the senses of the circles upon circles carrying the planets about, in addition to their reality in the real world of ideas. It would tell little unless one accepted the final cause, the actualization of the idea in the head of the architect, as explanation of the motion. Greek philosophers seemingly did accept such explanation.

Modern science has achieved a more elegant and informative solution to a more productively formulated problem, in the not-unbiased judgment of modern scientists.

The fact remains, though, that Plato and Eudoxus set for science a task accepted and pursued with varying vigor for two thousand years, from the fourth century B.C. through much of the Renaissance of the sixteenth and seventeenth centuries A.D.

Historical importance and impact are not necessarily commensurate with plausibility judged by different people in a different age.

Bibliography

On the change in science around 400 B.C., see F. M. Cornford's *Before and After Socrates*.

For Plato's view of reality and the true subject of the science of astronomy, see Cornford's *Plato's Republic*.

Excerpts from the *Republic* on the allegory of the cave and on the study of astronomy are adapted from translations by G. H. Wells, by J. L. Davies and D. J. Vaughan, and by Cornford.

For an introduction to Plato's cosmology, see *Plato's Timaeus* by Cornford, with text and comments.

Excerpts from the *Timaeus* are reprinted from Cornford's translation by permission of Humanities Press Inc., Atlantic Highlands, N.J. 07716.

Regarding Plato's place in the history of science, see G. E. R. Lloyd's article "Plato as a Natural Scientist" in the *Journal of Hellenic Studies, volume 88* (1968), pages 78-92.

For the tradition of Plato's opinions in astronomy, see Pierre Duhem's *To Save the Phenomena*, translated by E. Doland and C. Maschler.

Simplicius' comments on astronomy are reprinted from pages 5 and 23 of Duhem's book by permission of the University of Chicago Press.

The film *Plato's Cave* (20 minutes) dramatizes the allegory on human understanding.

Films on Socrates include: *The Greeks: In Search of Meaning* (26 minutes), in the *Western Civilization, Majesty and Madness* series; *Death of Socrates* (27 minutes), in the CBS *You Are There* series; and a different film (45 minutes) with the same title in the BBC *Plato* series, with a greater emphasis on philosophy.

On the general cultural background, see *Greece: The Golden Age* (29 minutes) and *Athens: The Golden Age* (29 minutes).

CHAPTER FIVE

Aristotle and Physical Astronomy

Plato's dictum that the appearances be saved by a system of uniform circular motions was one of the two major approaches to the study of astronomy set forth during the Hellenic period of the fourth century B.C. Astronomers working in the Platonic tradition centered their efforts upon devising a geometrical model. The model was a kinematic one: motions were described but their cause ignored.

Aristotle, in contrast to Plato, proposed a physical approach to the phenomena of nature. Astronomers or physicists later working in the Aristotelian tradition attempted to develop a dynamic model: the cause as well as a description of the motions was sought.

Major developments in astronomy in the Hellenistic period, during which the problems set in the preceding Hellenic period were pursued and answers developed, would follow primarily the pattern set by Plato. The physical or dynamical approach of Aristotle was not lost, however, and eventually both approaches were encompassed in astronomical studies.

Plato's philosophy can be understood as a response to his environment. Aristotle's very different philosophy may be interpreted in a similar manner, though with some difficulty, because for twenty years Aristotle was a member of Plato's Academy, from about 367 B.C. when Aristotle entered the Academy as a youth of 17 until Plato's death in 347 B.C.

The standard interpretation of Aristotle's work is that he began very close to Plato's intellectual position and only gradually departed from it. Ambiguities in the dating of Aristotle's writings encourage such an analysis, because the resulting pattern can then be used to determine the chronological order of the undated passages.

A different interpretation of Aristotle's work follows from thinking of his philosophy as governed by the interests of a biologist. This approach to an understanding of Aristotle fits well with biographical information and makes particularly good sense of Aristotle's writings on physics and astronomy. It is the interpretation that will be developed here. Of course, the convenience of any particular approach to Aristotle's philosophy is no guarantee of the accuracy of the following analysis!

Aristotle, born some forty-three years after Plato, in 384 B.C., experienced much more favorable and pleasant encounters with the world of the senses than had Plato. Consequently, Aristotle did not reject the world of the senses for the world of ideas as completely as had Plato.

Aristotle's father was the personal physician of Amyntas II of Macedon, a poor land of unruly people at the northern edge of the Greek peninsula. It is tempting to attribute to his physician father Aristotle's interest in the biological sciences, even though his father died when Aristotle was young.

At the age of seventeen, in 367 B.C., Aristotle left Macedon and journeyed south to Athens. At Plato's Academy, where he stayed for twenty years, Aristotle would have been introduced to the problems of identifying and classifying animals and explaining their functioning as part of nature. Admittedly, Aristotle also would have been exposed to all of Plato's philosophy, and the standard interpretation of Aristotle's philosophy is not implausible.

In 347 B.C. Plato died and Aristotle entered the second major period of his life. The situation in both Macedon and Athens had been changing. The feuding Greek city-states were ripe for conquest and Philip II, the successful claimant in 359 B.C. of the crown of Macedon, gradually consolidated his control. Athens emerged as his main opponent.

There was an anti-Macedonian mood in Athens. Coupled with it was Aristotle's dislike of the philosophical tendencies of Plato's nephew, who had taken over direction of the Academy following its founder's death. Thus was Aristotle encouraged to leave Athens.

His birthplace, Stagira, would have been the logical place to go, had it not been destroyed by Philip. Aristotle went instead across the Aegean Sea to Asia Minor.

There he founded a new academy under the patronage of a local ruler, Hermias, and married the ruler's eighteen-year old niece and adopted daughter. From Aristotle's later description of the ideal age for marrying as thirty-seven years for the man and eighteen for the woman, it may be inferred that his voluntary exile was not an unhappy one.

In 344 B.C. Aristotle went to the nearby island of Lesbos, where he worked with Theophrastus. Many of the observations in Aristotle's biological works were made in this region.

Aristotle returned to Macedon in 342 B.C. to tutor the young prince Alexander and, perhaps, to further secret negotiations between Hermias and Philip. Aristotle was not as concerned with speculations about the ideal state as Plato had been and was more actively involved in political activities. Many, but not all, of Aristotle's experiences were in successful causes. A noticeable exception was the capture of Aristotle's father-in-law, Hermias, by the Persians and his subsequent death during torture intended to elicit details of the secret treaty with Philip.

When Philip completed his conquest of Greece in 338 B.C., Aristotle was in the enviable position of helping form the ideas of the heir of the leading city-state of Greece. Alexander was more appreciative of Aristotle's efforts than had been the young Dionysius II of Plato's tutoring: Alexander rebuilt Stagira to please Aristotle.

In 336 B.C., following the assassination of his father, Alexander took the throne. Aristotle returned to Athens comfortably on the side of the victors.

Just before he died, Philip had planned a campaign against the Persian Empire. Alexander moved swiftly eastward. While conquering the Near East, he did not forget his tutor. Alexander sent a veritable flood of new plants and animals, as well as Babylonian eclipse records, back to Athens for Aristotle to study.

The political climate remained favorable for Aristotle in Athens until Alexander's death in 323 B.C. Aristotle then went into voluntary exile once more. He died a year later, in 322 B.C.

Typical of Aristotle's writings, and indeed of the biological sciences ever since, is an emphasis upon classification. Aristotle continually analyzed and classified, as if what was necessary to understand a subject was to divide it into categories.

In the *Poetics*, Aristotle divided poetry into tragedy and comedy, and analyzed tragedy into six factors: scenic presentment, lyrical song, diction, plot, character, and thought. In his *Politics*, Aristotle classified the species of polity, with their normal and perverted forms:

1) government of an individual - kingship and tyranny;
2) government of a few - aristocracy and oligarchy;
3) government of the many - polity and democracy.

In his scientific books, Aristotle took a similar approach. He classified and he drew distinctions. He distinguished between the mathematician and the natural philosopher, between the mathematical astronomy of Plato and his own dynamic astronomy.

Excerpt from Aristotle's *Physica*

Having distinguished the different meanings of "nature", let us see wherein the mathematician differs from the natural philosopher, for the solids, planes, lines, and points with which the mathematician deals are, after all, aspects of natural bodies; at the same time, let us consider more specifically whether or not astronomy is a part of natural science, for it would surely be unreasonable to expect the natural scientist to know what the sun and the moon are if he is not to know any of their essential attributes, especially when writers on nature patently do discuss what shape the sun and the moon have and whether or not the earth or the cosmos is spherical.

Now, although the mathematician, too, occupies himself with these things, he does not concern himself with them as limits of natural bodies or with the properties they have in this status; what he therefore does is to separate them, particularly since they can be separated in thought from processes without any resulting incongruity or falsity.

But the advocates of the "ideas" do this confusedly when in their theory they separate natural objects, which are much less separable than mathematical objects. This will become clear in any attempt to define these two kinds of entities and their respective attributes: odd and even, straight and curved, number, line, and figure can be independently of movement; not so things like flesh, bone, and man, which are defined like "snub nose"*, not like "curved".*

This difference is also clarified by reference to the more physical branches of mathematics (for example, optics, harmonics, and astronomy), which really proceed, in a sense, by a method the inverse of that of geometry: although geometry investigates physical lines, it ignores their physical aspects; optics, on the other hand, examines lines which are indeed mathematical, but deals with them not so much in their mathematical as in their physical aspects.

Accordingly, we must inquire into "nature", since it has two meanings: form and matter, in the same way in which we would consider what "snubness" is: reflection shows that such things are neither purely immaterial nor considered in their purely material aspect. So one may here, too, ask: With which of the two aspects of nature does the natural philosopher properly deal? Or does he deal with that composed of both of them together? If so, then he will also deal with each of them separately; but will he do this, then, in the same science or in separate sciences?

If we look for light on this problem to the ancients, natural philosophy would appear to be exclusively interested in matter, for even Empedocles and Democritus touched but lightly on questions of form and of what it means to be anything.

If, on the other hand, art imitates nature, and every scientific technique requires knowledge of its distinctive form as well as of the material suited to it - if, for example, a physician understands health as well as the bile and phlegm on which health depends, and a builder understands the form of a house as well as its materials (such as bricks and timber), and so forth, then natural philosophy must take cognizance of both [the formal and the material] aspects of nature.

One of the major strengths of Aristotle's philosophy was that it formed a complete system, a system in which every part seemed to follow logically from the other parts. To understand Aristotle's astronomy, it is necessary to understand his physics; basic concepts in Aristotle's *Physica*, or *Physics*, underly ideas developed in *De Caelo*, or *On the Heavens*. Aristotle's definitions of motion and place, his conception of what constitutes a cause, and his criteria for an acceptable answer to the question "Why?" are all essential to an understanding of his view of the physical universe.

To have knowledge of something, or to have grasped the "why" of it, was for Aristotle to know the cause of the phenomenon. Characteristically, Aristotle classified causes into categories:

1) the material cause - of what the object was made;
2) the formal cause - the shape of the object;
3) the efficient cause - who made it;
4) the final or purposeful cause - the object's use or purpose.

Excerpt from Aristotle's *Physica*

On the basis of these distinctions let us now examine what and how many sorts of explanatory factors there are. All inquiry aims at knowledge; but we cannot claim to know a subject matter until we have grasped the "why" of it, that is, its fundamental explanation. It must clearly, therefore, be our aim in the present inquiry to get knowledge of the first principles to which we may refer any problem in our exploration of generation and destruction and of any natural transformation.

"An explanatory factor", then, means (1) from one point of view, the material constituent from which a thing comes; for example, the bronze of a statue, the silver of a cup, and their kinds.

From another point of view, (2) the form or pattern of a thing, that is, the reason (and the kind of reason) which explains what it was to be that thing; for example, the factors in an octave are based on the ratio of two to one and, in general, on number. This kind of factor is found in the parts of a definition.

Again, (3) the agent whereby a change or a state of rest is first produced; for example, an adviser is "responsible" for a plan, a father "causes" his child, and, in general, any maker "causes" what he makes, and any agent causes what it changes.

Again, (4) the end or the where-for; so, when we take a walk for the sake of our health, and someone asks us why we are walking, we answer, "in order to be healthy", and thus we think we have explained our action. So any intermediate means to the end of a series of acts: for example, as means of health there are reducing, purging, drugs, instruments, and so forth; for all these are for an end, though they differ from one another in that some are instruments, and others are actions.

Since what we call an "explanatory factor" may be any one of these different aspects of a process, it follows not only that anything actually has several such factors which are not merely accidental differences of meaning (as both the sculptur's art and the bronze are needed to explain a statue as a statue, the bronze being its material, and the sculpturing its agent), but it follows also that these factors are reciprocal: for example, exercise explains good health, and good health explains exercises; though they explain each other differently (good health as end, and exercise as means). And the same thing may explain contraries: for the same thing which by its presence explains a given fact is "blamed" by its absence for the contrary fact; for example, a shipwreck is "caused" by the absence of the pilot, whose presence is responsible for the ship's safety.

All the factors here mentioned clearly fall under four varieties. From letters come syllables; from building materials come buildings; from fire, earth, and so forth, come bodies; from parts come wholes; and from assumptions come conclusions. The first factor in each of these pairs is the subject matter or the parts; the second is what it meant to be that particular whole, or synthesis, or form. A "cause" in the sense illustrated by a seed, a physician, an adviser, and any agent generally, is the factor whereby a change or state of being is initiated. Finally, there are the ends of the good of the others; for all the others tend toward what is best as toward their end. It makes no difference now whether we say "their good" or "their apparent good". ...

In explaining events, we often speak of "luck" and "chance", since many beings and happenings occur "by luck" or "by chance". Accordingly, we must ask: in what sense do luck and chance explain anything? Are they the same or different? In general, what is luck, and what is chance?

Some question whether there is any such thing. Nothing happens by luck, they say, but everything called "by luck" or "by chance" has some determinate explanation; ....

Again, some attribute the heavens and all worlds to chance happenings, on the theory that the "vortex" arose "of itself", that is, the motion which separated and arranged the entire universe in its present order.

This, too, is very "strange"! It is not by luck, they say, that animals and plants are and come to be but these are to be explained by nature or mind or something of the sort, since a seed does not develop into any random thing but one kind into an olive and another kind into a man, and so forth; yet the heavens, the most divine of all things visible, have no such determinate explanation as animals and plants have but arise spontaneously.

Even if this were the case, it would be a situation that might well give us pause and prompt us to say something of importance about it. Apart from the strangeness of the theory in other respects, it is especially "strange" to assert chance in the case of the heavens where we do not observe it and to deny luck to a region in which many events occur by luck; what happens is the very opposite of what their theory would lead us to expect. ...

Clearly there are explanatory factors, and they fall into four types corresponding, as we have analyzed them, to the four meanings of the question "why". The "why" of anything may ultimately mean: (1) "what it is", for example, a straight line or the commensurable or anything else as defined in mathematics or in the realm of unchanging considerations generally; or (2) "what started a process", for example, a war, as in revenge for a raid; or (3) "to what end", for example, in order to win rulership; or (4) in the case of products, their "material". Since these are the four types of explanation, the natural philosopher must try to understand them all if he is to deal adequately with the "why" of anything in terms of each type of explanatory factor: the material, the form, the agent, the "where-for".

Often the first three factors coincide. "What" something is and "to what end" it is, may be the same; and a prime mover may be identical with these factors in form or species, since man generates man, and so with moved movers generally. As for unmoved movers, they do not belong to the domain of natural science, since they evoke movement without having within themselves either movement or a beginning of movement. In this regard, therefore, we must distinguish three sciences: one, concerning unmoved beings; another, concerning moved beings which are imperishable; and a third, concerning moved beings which are perishable.

Aristotle frequently explained a thing in terms of its purpose. For example, he explained that a man is walking because the man wants to be healthy. Aristotle thus shared with Plato a tendency to use teleological explanations.

Aristotle's is also a very animistic science. He often placed the source of motion within an object itself rather than explain motion in terms of outside forces. To the extent that Aristotle seems content in his search for causes merely to restate the effects as undefined powers producing the effects, his scientific methodology has been strongly criticized.

The emphasis on the final cause - on the purpose of an action - helps to explain Aristotle's otherwise seemingly strange definition of motion. Motion, for Aristotle, was not solely change of position; it was more broadly defined as the fulfillment of potentiality. Motion thus included the growth of a plant or animal as it strove to attain its form, as well as what is now normally considered motion: a change of position - or locomotion, as it was called by Aristotle. Both types of motion involved the concept of potentiality and purposeful action.

Excerpt from Aristotle's *Physica*

Since we are exploring nature as a beginning of movement and change, we must not neglect to define what movement is, lest unclearness about movement lead us into inevitable unclearness about nature; and then, having defined movement, we must in the same manner take up the terms which follow from it. . . .

Since any kind of being may be distinguished as either potential or completely realized, the functioning of what is potential as potential, that is "being in movement": thus, the functioning of the alterable as alterable is "qualitative alteration"; the functioning of that which can *increase or decrease the contrary (there is no name common to both) is "increase" or "decrease"; the functioning of that which* can *be generated or destroyed is "generation" or "destruction"; the functioning of what* can *change its place is local motion. . . .*

The functioning, therefore, of what is potential, when there is actually something being realized, not as itself, but as movable, is motion. . . .

The Aristotelian sense of motion leads to a particular sense of place. The Aristotelian doctrine of place encompasses both motion and potential. Each of the four elements - earth, water, air, and fire - has its natural place. Moved away from its natural place, each element has a natural tendency to return to its natural place. Thus, to explain a motion, one could attribute the motion to a final cause or purpose: the object is moving because it has a tendency to return to its proper place in the universe. Fire moves upward toward its natural place and earth falls downward toward its natural place. Earth can be thrown upward with an unnatural motion.

Excerpt from Aristotle's *Physica*

A natural scientist must inform himself not only on the infinite but also on place. Is there such a thing or not; and if so, how is it, and what is it? People generally suppose existing things to be somewhere and nonexisting things to be nowhere, for where would a goat-stag or a sphinx be? Moreover, the kind of "movement" involved in all the types of change and most strictly so called is change of place or (as we are accustomed to designating it) "local motion". However, the question "What is place?" is fraught with many difficulties. All the facts appear to give rise to divergent theories. Besides, nothing has come down to us from our predecessors either by way of a statement or by way of a solution of the problems concerning place.

Nevertheless, it clearly seems to be a fact that place "is". First, there is displacement. Where *now there is water,* there *will be air when the water has gone (as out of a vessel); and then again some other body will occupy the same place. The place, therefore, seems to be different from all the bodies which successively displace one another. That "in" which the air is now, is that "in" which the water was before. Consequently, the place was clearly something; that is, the location was different from the bodies which, by passing into and out of it, changed places.*

Secondly, the motions of the simple bodies (fire, earth, and so forth) show not only that place is something but also that place has some kind of functional significance. Unless interfered with, each of the simple bodies moves up or down to its appropriate place. The six directions, up and down and so forth [right and left, before and behind], are parts or kinds of places; and they are such not only relatively to ourselves, but they are up and down, right and left, and so forth. For us, they do not always remain the same but change as we change our position; so that often even the same thing may be both right and left, both above and below, both before and behind. But in nature each of these is distinct independently of our own position. Not any chance direction is "up", but that in which fire and light bodies move; so, too, not any chance direction is "down", but that in which heavy and earthy bodies move. Thus, these distinctions of place do not only differ conventionally but also depend upon the ways in which bodies act. . . .

This is the reason why the center of the cosmos and the inner surface of the rotating celestial system are conceived as functionally down and up for all men, because the former is always stationary and the latter remains coincident with itself. Since light bodies naturally tend upwards and heavy bodies downwards, the surrounding boundary towards the center of the cosmos and the center itself are "down", whereas the boundary towards the outermost part of the cosmos and the outermost part itself are "up".

Aristotle's concept of place led him to the conclusion that void could not exist, for if it did, absurd conclusions could be drawn from the fact of its existence. This type of argument is termed *reductio ad absurdum.* To argue in this fashion is to start with a seemingly plausible statement and then deduce such absurd consequences from it that one is forced to conclude that the original statement obviously cannot be true. Thus did Aristotle argue against the existence of a void, and develop some important views on motion.

Excerpt from Aristotle's *Physica*

Let us continue our argument against those who ascribe independent existence to a "void".

If each of the simple bodies naturally moves in a definite direction (fire upward, and earth downward and towards the middle of the universe), it is clear that a void cannot explain local motion: since the void is thought to explain local motion but not any of those mentioned, which local motion, then, does it explain?

Again, if a "void" is a sort of "place deprived of body", where would a body placed in a void move to? Surely not into the whole of the void! The same argument applies here that was used against the conception of a place as an independent something into which things are transported: how would anything placed in it start moving or stop moving? The same argument is no less applicable to "up" and "down" and to the "void" - for those who maintain the existence of a void represent it as a place. Then, too, how can anything be "in" a place or a void? This does not happen when a whole body is located in a supposedly severed place and abiding body: any part not separately located there would not be in a place, but in the whole [of which it is a part]! Then, again, if a place has no [segregated] existence, neither does a void.

Indeed, as against those who say that a void is necessary if there is to be movement, reflection shows the very opposite: were there an independently existing void, there would not be a single movement! Just as the earth is stationary (according to the arguments of some) because of the uniformity of its medium, so, too, in a void [sharply demarcated from bodies] there would have to be complete stability: since there would be no differences there, neither would there be any place for anything to move to rather than to some other place.

Another reason [in support of our contention] is the difference between violent and natural motion: being contrary to natural motion, violent motion presupposes the latter in the sense that if there is violent motion there must also be natural motion; and therefore, if the various natural bodies do not have their natural motions, they will not be capable of violent motion either. But how can there be natural motions in a void or in an infinite in which there are no differences? No more than an infinite as such has any up or down or middle, does a void as such have any difference of up or down: just as there are no differences in nothing, so there can be none in nonbeing, and the void seems to be a sort of nonbeing or privation. But natural motions are differentiated; and their distinctions [of up and down] are natural. Either, then, nothing has a natural motion, or else there is no void!

Again, projectiles continue to move even after what has propelled them is no longer in contact with them. Some explain this phenomenon by mutual replacement. Another explanation may be that the air which has been pushed pushes projectiles with a motion more vigorous than their motion to their resident place. But none of these things can happen in a void: there, a body can continue moving only as long as it is propelled by something else.

Again, it would be impossible to state any reason why anything set in motion would stop anywhere. Why would it stop at one place rather than at another? Hence, a body would either continue in its state of rest or would necessarily continue in its motion indefinitely, unless interfered with by a stronger force.

Again, things are held to move into a void because of its yielding character. However, a void would yield equally everywhere; so that any motion in it would be in every direction.

There are further considerations which make what we have said even more evident. We see a body of a certain weight moving at a faster or lower rate for one of two reasons: either, on the one hand, because of a difference in the medium, for example, in earth or water or air; or, on the other hand, because of a difference in the moving bodies compared such that, other things being equal, one body is heavier or lighter than the other.

On the one hand, then, the medium explains [different velocities] because of its resistance; especially when the medium is moving in an opposite direction, but also even when the medium is in a state of rest. This is especially true of a medium which is not easily divided or which, in other words, is relatively dense.

Thus, a body (B) will move through a medium such as water (W) in a given time (Tw) and through a thinner medium such as air (A) in a different time (Ta), namely, in proportion to the density of the hindering body (provided W and A are equal in length): the thinner and less solid the air than the water, the faster will B move through A than through W; one speed has the same ratio to the other that the air has to the water.

Hence, if the air is twice as thin as the water, the body will move through W in twice the time it takes to move through A; and the time Tw will be twice the time Ta. Likewise, always, the less solid and resistant and the more readily divided the medium, the faster will be the motion.

There is no definite ratio, however, in which a void stands to a body. So zero is in no ratio to a number: although 4 exceeds 3 by 1, and 2 by more than 1, and 1 by still more, there is no ratio by which 4 is greater than 0; but inasmuch as the greater has to be made up of the less and the remainder, 4 would consist of 0 and that by which 4 is greater than 0. So, too, a line is not "greater than" a point, unless a line were composed of points. By the same token, therefore, as there is no ratio between the "empty" and the "full", so there is no ratio between movement [through the former and through the latter]; on the contrary, if a thing moves through the thinnest medium over a certain distance in a certain time, then it will move through the void with a velocity beyond any ratio!

Let V be a void equal in magnitude to W and to A. Then if B is to move through V in a time Tv which is less than time Ta, the "void" would be in that ratio to the "full".

But in a time equal to Tv, B would move through a part P of the medium A. Also, in time Tv, B would span any body V which is thinner than air by as much as time Ta is longer than time Tv; for if the body V is as much thinner than A as Ta exceeds Tv, B would move through V (if at all) in a time inverse to the speed, that is, in a time equal to Tv.

If V is empty, then, B will pass through it still faster. But the assumption was that B passed through V when V was empty in time Tv.

Consequently, B would move through V in an equal time regardless of whether V was full or empty. But this is impossible.

Evidently, then, if there is a time in which anything moves through a part of a void, this impossible result will follow: a body will move over a certain distance, full or empty, in equal times, since there will be some body bearing to the other body the same ratio which the one time bears to the other time. In short, the reason for this result is clear: any motion stands to another in a certain ratio, since the motions occur in time and there is a ratio between any finite times; but there is no ratio [of density] between anything that is empty and anything that is full.

On the other hand, having considered what follows from the differences among the media, let us turn to consider the difference between one moving body and another, together with its consequences.

We see that, other things being equal, heavy and light bodies move with unequal velocities over an equal space in the ratio which their magnitudes have to each other.

Hence, they must do so even when moving through a void. But this is impossible. Why should one of them move faster than another? To be sure, one of them does necessarily move faster in a [space] full [of matter] because the greater body divides the medium faster by reason of its force; for a body moving naturally or a projectile divides a medium as it does by reason either of its shape or of its [upward or downward] tendency. But [in a void] everything would, accordingly, move with equal velocity. But this is impossible.

Aristotle's physics was closely woven; no part could be separated from the rest; no part could be changed without changing the whole. The concept of place, upon which depended an explanation of motion - bodies returning to their natural place - was not applicable to a void, with no place; thus the laws of natural motion would not work in a void; hence there could not be a void, because if there were, Aristotle's physics would not work. Put another way, Aristotle's physics denied the existence of a void to maintain its logical completeness.

Nor was projectile motion explicable in a void. Movement required constant contact between the moved object and the mover. But in a void there was only the moved object. There was no air, and no possibility of explaining projectile motion in terms of air pushing the object.

There was another problem with motion in a void. What reason was there for the motion to cease, ever, if there were nothing to stop it?

Modern physics, of course, does contemplate a body remaining at rest or in motion until acted upon by another force. But that goes against common sense and observations. Friction is sufficient to eliminate perpetual motion from the real world, except for astronauts in outer space. Aristotle had not the advantage of this perspective, and perpetual motion seemed absurd to him. That perpetual motion was a logical consequence of void was sufficient reason to reject the possibility of a void.

Also arguing against the possibility of a void was Aristotle's concept of the ratio of speeds of equal objects being in proportion to the thickness of the medium. If water were twice as thick as air, an object would move through water with half the speed it moved through air. But void with no thickness made such a ratio nonsense; it meant dividing by zero.

Furthermore, for falling objects, which Aristotle believed fell at different speeds depending upon their weight, or upon how much force they had to cut the medium, fall in a void would produce the patently nonsensical result of bodies of different weight falling with the same speed because there was no medium to slow one more than the other. Indeed, this is exactly what modern physics says will happen! But in Aristotle's time it seemed absurd, and if the logical consequence of a void was absurd, then the concept of void itself was absurd, too.

The case of a thrown object, of projectile motion, raised a problem which would puzzle and plague generations of philosophers and scientists after Aristotle. Eventually, failure to solve the problem helped lead to the downfall of the entire Aristotelian physical theory.

Everything in locomotion - movement from one place to another - either moved itself or was moved by something else. And, according to Aristotle, it was impossible to move something without being in continuous contact with it.

The insistence on continual contact between the moved and the mover made it difficult to explain projectile motion. In the above passage from the *Physica*, Aristotle briefly mentions projectiles, attributing their continued motion either to mutual replacement or to the air being pushed by the thrower in turn pushing the projectile. By mutual replacement, Aristotle meant the doctrine of antiperistasis, in which the air pushed forward by the projectile changes direction and moves around to the rear of the projectile and then, changing direction once more, gives the projectile a push forward.

This explanation was not especially convincing, nor was that of the air being pushed by the thrower of the projectile and in turn pushing the projectile. Eventually attempts to explain projectile motion would lead to the concepts of impetus and momentum and on to the concept of inertia and a body remaining in motion until some force acted to stop it.

In Aristotle's time, though, and for centuries after, it was necessary to explain forced or violent motion as the result of continuous contact between the moved and the mover. When the contact ceased, so supposedly would the motion. The opposite was natural motion, in which a body moved naturally toward its natural place in the universe without the interposition of any force other than the natural tendency to move to the natural place.

Excerpt from Aristotle's *Physica*

Now, of the things which are moved essentially, some are moved by themselves, whereas others are moved by something else; and some are moved naturally, whereas others are moved "contrary to nature" or violently.

Anything moved by itself (such as any animal) is moved naturally; for when an animal is moved by itself inasmuch as it has within itself the beginning of its movement, we say that its movement is "natural". But although an animal as a whole thus moves itself naturally, its body may be moved naturally or violently; this depends upon the kind of movement it happens to be undergoing or upon the kind of element which may enter into its composition.

Anything moved by something else is moved naturally or violently. Examples of violent movement include the upward motion of earthy things and the downward motion of fire as well as such motions of animal parts as are at variance with their normal position or manner. Violent movements afford the clearest evidence that whatever is moved is moved by an agent.

Next in order, this circumstance is manifest in those natural movements which are exemplified by animals moving themselves. What may be obscure in such movements is not that they are effected by an agent, but rather how we are to distinguish in them between the mover and the thing moved. It would seem that a mover and a thing moved are to be distinguished in movements of animals no less than in movements of ships or of any artifacts and that, moreover, it is only in this sense that an animal as a whole is a self-moving agent.

But the distinction in question is made with the greatest difficulty in the class of things remaining according to our classification. We have noted, among things moved by something else, those which are moved violently. But there remains, among things moved by something else, a class of things opposed to those which are moved violently; namely, the class of things moved naturally.

Things whose movements are not only brought about by something else, but are also natural to them, confront us with the question: by what are they moved? Consider, for example, things light and heavy. Evidently, they are moved in directions respectively alien to themselves by main force; but they are naturally moved to their respective resident places, that is, light things upward and heavy things downward. In their natural movements, however, it is no longer equally evident by what agency their movement is effected. Surely, they cannot set themselves in motion, any more than they can bring themselves to a stop, as animals can. If fire, for example, could of itself move up as an animal can of itself start walking, then fire could presumably also of itself move downward as an animal can of itself stop walking. It would, moreover, be unreasonable to suppose that, if [elements] moved of their own accord, they would be capable of only a single kind of self-initiated "movement". Besides, how would it be possible for anything continuous and naturally unitary to bestir itself? In so far as anything is one and continuous (rather than merely contiguous), in so far it is incapable of being acted upon. Things must admit of division if they are to function both actively and passively. Hence, nothing naturally unitary [like fire, and so forth] can be self-moving, any more than can anything else that is continuous; but within such things, the active and the passive factors must be just as distinguishable as we observe them to be in the action of living things upon nonliving things.

Now, nonliving things are [even in their natural movements] moved by something. This becomes evident if we distinguish different factors accounting for their movements; that is, if we divide movers, as we have divided things moved, into those whose movements are violent and those whose movements are natural to them. Thus, a lever moves a weight by main force, whereas a body which is actually hot can act in accordance with this very nature upon a body which is potentially hot; and so forth. Just so, a body can be acted upon in accordance with its nature if it is potentially of a certain sort or so much or somewhere; provided that the process in question can arise in it, and provided that the process would not be incidental to it. (Although the same thing may be [potentially] of some character as well as of some dimension, still, the one aspect would not belong to the other essentially but would be incidental to it.) Fire and earth, then, are forcibly moved by some agent in ways at variance with their nature; they are moved naturally when they are impelled to pass from their respective potential states to their corresponding activities.

The "potential", however, has more than one meaning. This is the reason why the agent may escape observation in the upward motion of fire and the downward motion of earth. Consider an analogy: a learner is a potential knower at one stage; he becomes a potential knower at another stage when he has acquired knowledge which, at a given time, he is not putting to actual use. In any case, however, the interaction of something capable of acting with somthing capable of being acted upon is the occasion for a passage from a potentiality to an actuality. Thus, a learner passes from his [first] potential knowing to a second potential knowing: from the knowing which is potential in the sense that the learning process has not been completed, to the differently potential knowing of one who does not turn the knowledge he possesses to actual account. But one who has arrived at the latter stage puts into practice his intelligence, unless his activity is impeded; if he did not do so, he would really be in the opposite state of ignorance. So it is with natural bodies [or elements]: a cold body is potentially hot; when turned into fire, it burns - unless it is prevented from doing so. So, too, with heavy and light bodies: when water (the first heavy body which is potentially light) is transformed into air, it becomes a light body and at once acts as such (unless hindered); for light bodies tend to rise and, when held "down", are prevented from doing so. A similar analysis may be made of quantitative and qualitative changes, with similar results.

We may, however, persist in asking: why, after all, do heavy and light bodies tend to their own respective places? The reason is that they naturally tend in distinct directions. This is precisely what it means for them to be light or heavy; namely, that they tend upward or downward, respectively. But, as we have said, there is more than one way of being potentially light or potentially heavy. When a thing is water, it is at least in some sense potentially light; and when it is air, it is still only potentially light in the sense that its upward tendency may be counteracted; whereas, on the removal of the obstacle, it becomes actually light and proceeds to rise. Analogously, what is [potentially] of a certain quality becomes actually so, as when an expert is at liberty to think. Likewise, what is of a certain quantity expands when unchecked. As for anyone who removes an obstruction such as a pillar from under a roof or a weight from a wine-skin in a body of water, he in a way moves the object he has freed; that is, he moves it incidentally only, somewhat as a wall intervenes to cause a ball to rebound which a thrower has hurled against it. Clearly, then, none of the things we are talking about move themselves; yet they have a "source of movement". Not that they produce any movement or keep it going; but they are amenable to movement.

In fine, the movements of things in their courses are natural or violent; and the latter are due to external agents, whereas the former may be enacted by self-moving agents or be due to external agents. Thus, light and heavy things have two sorts of "movers": those which have produced them and made them light and heavy; and those which have freed them from the hindrances to their movements. Hence, all things in movement are acted upon by some agent.

In the last book of the *Physica,* Aristotle discusses the one form of locomotion that can be continuous. Locomotion is either rotatory or rectilinear or a combination of the two, and it is only rotatory motion that can be continuous. Furthermore, rotation is the primary locomotion because it is more simple and complete than rectilinear motion. This value judgment regarding the relative merits of circular and rectilinear motion was to play an important part in the subsequent history of astronomy.

Excerpt from Aristotle's *Physica*

Clearly, circular motion is the primary form of local motion. We have divided local motion into the circular, rectilinear, and composite. But the simple kinds must have priority over the composite, which is made up of them; and the circular must have priority over the rectilinear, than which it is the simpler and more complete. Motion along an infinite straight line is impossible, since nothing is infinite in this way; and if there were an infinite of this sort, nothing could span it (since the impossible does not happen, and it is impossible for the infinite to be spanned). And motion to and fro along a finite line is a compound of two motions, whereas a single motion along a finite line is incomplete and soon at an end. But the complete is prior to the incomplete, and the imperishable is prior to the perishable, in the order of nature and of definition and of time. Then, too, a motion which can be eternal has priority over one which cannot. And circular motion can be eternal, as no other motion or change can; in the latter, pauses must occur at which movement ceases.

Aristotle's views on the organization and structure of the universe are presented in his *De Caelo*, or *On Heavenly Bodies.* This work contains some of the basic considerations of motion also present in the *Physica.* All locomotion is either straight, circular, or a combination of the two, and all bodies are either simple - that is, composed of a single element, such as fire or earth - or are compounds. The element fire and bodies composed of it have a natural movement upward; bodies composed of earth have a natural movement downward (actually, toward the center of the universe, and the earth is thus at the center). Circular movement is natural to some substance other than the four elements - earth, fire, air, and water - and more divine than these four elements (since circular motion is prior to straight movement).

Excerpt from Aristotle's *De Caelo*

The question as to the nature of the whole, whether it is infinite in size or limited in its total mass, is a matter for subsequent inquiry. We will now speak of those parts of the whole which are specifically distinct.

Let us take this as our starting-point. All natural bodies and magnitudes we hold to be, as such, capable of locomotion; for nature, we say, is their principle of movement.

But all movement that is in place, all locomotion, as we term it, is either straight or circular or a combination of these two, which are the only simple movements. And the reason of this is that these two, the straight and the circular line, are the only simple magnitudes. Now revolution about the center is circular motion, while the upward and downward movements are in a straight line, "upward" meaning motion away from the center, and "downward" motion towards it. All simple motion, then, must be motion either away from or towards or about the center.

This seems to be in exact accord with what we said above: as body found its completion in three dimensions, so its movement completes itself in three forms. Bodies are either simple or compounded of such; and by simple bodies I mean those which possess a principle of movement in their own nature, such as fire and earth with their kinds, and whatever is akin to them.

Necessarily, then, movements also will be either simple or in some sort compound - simple in the case of the simple bodies, compound in that of the composite - and in the latter case the motion will be that of the simple body which prevails in the composition.

Supposing, then, that there is such a thing as simple movement, and that circular movement is an instance of it, and that both movement of a simple body is simple and simple movement is of a simple body (for if it is movement of a compound it will be in virtue of a prevailing, simple element), then there must necessarily be some simple body which revolves naturally and in virtue of its own nature with a circular movement.

By constraint, of course, it may be brought to move with the motion of something else different from itself, but it cannot so move naturally, since there is one sort of movement natural to each of the simple bodies.

Again, if the unnatural movement is the contrary of the natural and a thing can have no more than one contrary, it will follow that circular movement, being a simple motion, must be unnatural, if it is not natural, to the body moved. If then (1) the body, whose movement is circular, is fire or some other element, its natural motion must be the contrary of the circular motion. But a single thing has a single contrary; and upward and downward motion are the contraries of one another. If, on the other hand, (2) the body moving with this circular motion which is unnatural to it is something different from the elements, there will be some other motion which is natural to it. But this cannot be. For if the natural motion is upward, it will be fire or air, and if downward, water or earth.

Further, this circular motion is necessarily primary. For the perfect is naturally prior to the imperfect, and the circle is a perfect thing. This cannot be said of any straight line: - not of an infinite line; for, if it were perfect, it would have a limit and an end: nor of any finite line; for in every case there is something beyond it, since any finite line can be extended.

And so, since the prior movement belongs to the body which is naturally prior, and circular movement is prior to straight, and movement in a straight line belongs to simple bodies - fire moving straight upward and earthy bodies straight downward towards the center - since this is so, it follows that circular movement also must be the movement of some simple body. For the movement of composite bodies is, as we said, determined by that simple body which preponderates in the composition. These premises clearly give the conclusion that there is in nature some bodily substance other than the formations we know, prior to them all and more divine than they.

But it may also be proved as follows. We may take it that all movement is either natural or unnatural, and that the movement which is unnatural to one body is natural to another - as, for instance, is the case with the upward and downward movements, which are natural and unnatural to fire and earth respectively. It necessarily follows that circular movement, being unnatural to these bodies, is the natural movement of some other. Further, if, on the one hand, circular movement is natural *to something, it must surely be some simple and primary body which is ordained to move with a natural circular motion, as fire is ordained to fly up and earth down. If, on the other hand, the movement of the rotating bodies about the center is* unnatural, *it would be remarkable and indeed quite inconceivable that this movement alone should be continuous and eternal, being nevertheless contrary to nature.*

At any rate the evidence of all other cases goes to show that it is the unnatural which quickest passes away. And so, if, as some say, the body so moved is fire, this movement is just as unnatural to it as downward movement; for any one can see that fire moves in a straight line away from the center.

On all these grounds, therefore, we may infer with confidence that there is something beyond the bodies that are about us on this earth, different and separate from them; and that the superior glory of its nature is proportionate to its distance from this world of ours.

Aristotle thus infers that there is something beyond the region of the earth, composed of a different material, of a superior glory to our region of the earth, and also unalterable. His model is conveniently described as a two-sphere universe, with the changing region up to the sphere of the moon, the earth in the center surrounded by water, with air and fire at the top, beyond which are the heavenly bodies in circular motion and in a realm without change. There is a separate set of physical laws for each of the two regions, since they are composed of different types of matter.

Excerpt from Aristotle's *De Caelo*

In consequence of what has been said, in part by way of assumption and in part by way of proof, it is clear that not every body either possesses lightness or heaviness.

As a preliminary we must explain in what sense we are using the words "heavy" and "light", sufficiently, at least, for our present purpose: we can examine the terms more closely later, when we come to consider their essential nature.

Let us then apply the term "heavy" to that which naturally moves towards the center, and "light" to that which moves naturally away from the center. The heaviest thing will be that which sinks to the bottom of all things that move downward, and the lightest that which rises to the surface of everything that moves upward. Now, necessarily, everything which moves either up or down possesses lightness or heaviness or both - but not both relatively to the same thing: for things are heavy and light relatively to one another; air, for instance, is light relatively to water, and water light relatively to earth.

The body, then, which moves in a circle cannot possibly possess either heaviness or lightness. For neither naturally nor unnaturally can it move either towards or away from the center. Movement in a straight line certainly does not belong to it naturally, *since one sort of movement is, as we saw, appropriate to each simple body, and so we should be compelled to identify it with one of the bodies which move in this way. Suppose, then, that the movement is* unnatural. *In that case, if it is the downward movement which is unnatural, the upward movement will be natural; and if it is the upward which is unnatural, the downward will be natural. For we decided that of contrary movements, if the one is unnatural to anything, the other will be natural to it. But since the natural movement of the whole and of its part - of earth, for instance, as a whole and of a small clod - have one and the same direction, it results, in the first place, that this body can possess no lightness or heaviness at all (for that would mean that it could move by its own nature either from or towards the center, which, as we know, is impossible); and, secondly, that it cannot possibly move in the way of locomotion by being forced violently aside in an upward or downward direction. For neither naturally nor unnaturally can it move with any other motion but its own, either itself or any part of it, since the reasoning which applies to the whole applies also to the part.*

It is equally reasonable to assume that this body will be ungenerated and indestructible and exempt from increase and alteration, since everything that comes to be comes into being from its contrary and in some substrate, and passses away likewise in a substrate by the action of the contrary into the contrary, as we explained in our opening discussions. Now the motions of contraries are contrary. If then this body can have no contrary, because there can be no contrary motion to the circular, nature seems justly to have exempted from contraries the body which was to be ungenerated and indestructible, for it is in contraries that generation and decay subsist.

Again, that which is subject to increase increases upon contact with a kindred body, which is resolved into its matter. But there is nothing out of which this body can have been generated. And if it is exempt from increase and diminution, the same reasoning leads us to suppose that it is also unalterable. For alteration is movement in respect of quality; and qualitative states and dispositions, such as health and disease, do not come into being without changes of properties. But all natural bodies which change their properties we see to be subject without exception to increase and diminution. This is the case, for instance, with the bodies of animals and their parts and with vegetable bodies, and similarly also with those of the elements. And so, if the body which moves with a circular motion cannot admit of increase or diminution, it is reasonable to suppose that it is also unalterable.

The reasons why the primary body is eternal and not subject to increase or diminution, but unaging and unalterable and unmodified, will be clear from what has been said to any one who believed in our assumptions.

For all men have some conception of the nature of the gods, and all who believe in the existence of gods at all, whether barbarian or Greek, agree in allotting the highest place to the deity, surely because they suppose that immortal is linked with immortal and regard any other supposition as inconceivable. If then there is, as there certainly is, anything divine, what we have just said about the primary bodily substance was well said.

The mere evidence of the senses is enough to convince us of this, at least with human certainty. For in the whole range of time past, so far as our inherited records reach, no change appears to have taken place either in the whole scheme of the outermost heaven or in any of its proper parts.

The common name, too, which has been handed down from our distant ancestors even to our own day, seems to show that they conceived of it in the fashion which we have been expressing. The same ideas, one must believe, recur in men's minds not once or twice but again and again. And so, implying that the primary body is something else beyond earth, fire, air, and water, they gave the highest place a name of its own, aither, derived from the fact that it "runs always" for an eternity of time. Anaxagoras, however, scandalously misuses this name, taking aither as equivalent to fire.

Aristotle's universe is not infinite, as he argued, because the universe moves in a circle (as we can see with our eyes if we watch the stars). If the universe were infinite, then it would be moving through an infinite distance in a finite time, which is impossible. Thus by beginning with that which he disputed, that the world is infinite, and showing that the premise leads to an absurd conclusion, Aristotle, by the method of *reductio ad absurdum*, proved the opposite.

Excerpt from Aristotle's *De Caelo*

This being clear, we must go on to consider the questions which remain. First, is there an infinite body, as the majority of the ancient philosophers thought, or is this an impossibility? The decision of this question, either way, is not unimportant, but rather all-important, to our search for the truth. It is this problem which has practically always been the source of the differences of those who have written about nature as a whole. So it has been and so it must be; since the least initial deviation from the truth is multiplied later a thousandfold. Admit, for instance, the existence of a minimum magnitude, and you will find that the minimum which you have introduced, small as it is, causes the greatest truth of mathematics to totter. The reason is that a principle is great rather in power than in extent; hence that which was small at the start turns out a giant at the end. Now the conception of the infinite possesses this power of principles, and indeed in the sphere of quantity possesses it in a higher degree than any other conception; so that it is in no way absurd or unreasonable that the assumption that an infinite body exists should be of peculiar moment to our inquiry. The infinite, then we must now discuss, opening the whole matter from the beginning.

Every body is necessarily to be classed either as simple or as composite; the infinite body, therefore, will be either simple or composite. But it is clear, further, that if the simple bodies are finite, the composite must also be finite, since that which is composed of bodies finite both in number and in magnitude is itself finite in respect of number and magnitude: its quantity is in fact the same as that of the bodies which compose it. What remains for us to consider, then, is whether any of the simple bodies can be infinite in magnitude, or whether this is impossible. Let us try the primary body first, and then go on to consider the others.

The body which moves in a circle must necessarily be finite in every respect, for the following reasons. (1) If the body so moving is infinite, the radii drawn from the center will be infinite. But the space between infinite radii is infinite: and by the space between the radii I mean the area outside which no magnitude which is in contact with the two lines can be conceived as falling. This, I say, will be infinite: first, because in the case of finite radii it is always finite; and secondly, because in it one can always go on to a width greater than any given width; thus the reasoning which forces us to believe in infinite number, because there is no maximum, applies also to the space between the radii. Now the infinite cannot be traversed, and if the body is infinite the interval between the radii is necessarily infinite: circular motion therefore is an impossibility. Yet our eyes tell us that the heavens revolve in a circle, and by argument also we have determined that there is something to which circular movement belongs.

(2) Again, if from a finite time a finite time be subtracted, what remains must be finite and have a beginning. And if the time of a journey has a beginning, there must be a beginning also of the movement, and consequently also of the distance traversed. This applies universally. Take a line, ACE, infinite in one direction, E, and another line BB, infinite in both directions. Let ACE describe a circle revolving upon C as center. In its movement it will cut BB continuously for a certain time. This will be a finite time, since the total time is finite in which the heavens complete their circular orbit, and consequently the time subtracted from it, during

which the one line in its motion cuts the other, is also finite. Therefore there will be a point at which ACE began for the first time to cut BB. This, however, is impossible. The infinite, then cannot revolve in a circle; nor could the world, if it were infinite.

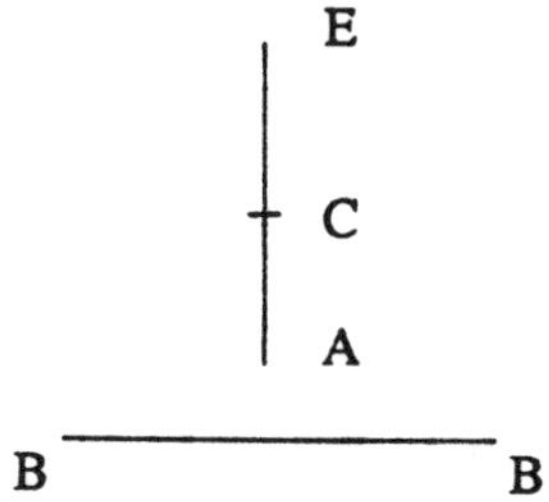

(3) That the infinite cannot move may also be shown as follows. Let A be a finite line moving past the finite line B. Of necessity A will pass clear of B and B of A at the same moment; for each overlaps the other to precisely the same extent. Now if the two were both moving, and moving in contrary directions they would pass clear of one another more rapidly; if one were still and the other moving past it, less rapidly; provided that the speed of the latter were the same in both cases. This, however, is clear: that it is impossible to traverse an infinite line in a finite time. Infinite time, then, would be required. . . . If, therefore, the time which the finite moving line takes to pass the other is infinite, then necessarily the time occupied by the motion of the infinite past the finite is also infinite. For the infinite to move at all is thus absolutely impossible; since the very smallest movement conceivable must take an infinity of time. Moreover the heavens certainly revolve, and they complete their circular orbit in a finite time; so that they pass round the whole extent of any line within their orbit, such as the finite line AB. The revolving body, therefore, cannot be infinite.

(4) Again, as a line which has a limit cannot be infinite, or, if it is infinite, is so only in length, so a surface cannot be infinite in that respect in which it has a limit; or, indeed, if it is completely determinate, in any respect whatever. Whether it be a square or a circle or a sphere, it cannot be infinite, any more than a foot-rule can. There is then no such thing as an infinite sphere or square or circle, and where there is no circle there can be no circular movement, and similarly where there is no infinite at all there can be no infinite movement; and from this it follows that, an infinite circle being itself an impossibility, there can be no circular motion of an infinite body.

(5) Again, take a center C, an infinite line AB, another infinite line at right angles to it, E, and a moving radius CD. CD will never cease contact with E, but the position will always be something like CE, CD cutting E at F. The infinite line, therefore, refuses to complete the circle.

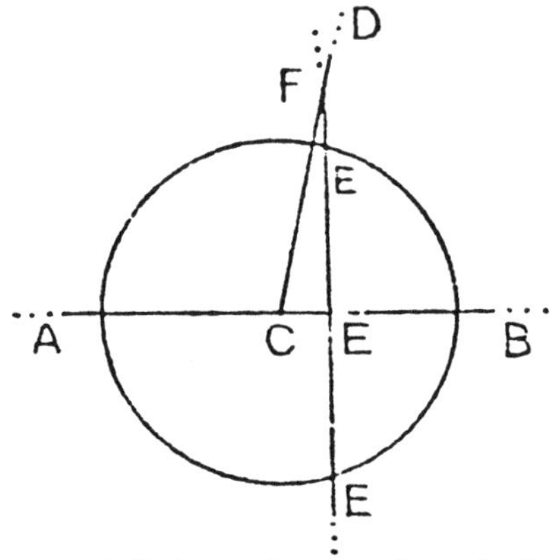

(6) Again, if the heaven is infinite and moves in a circle, we shall have to admit that in a finite time it has traversed the infinite. For suppose the fixed heaven infinite, and that which moves within it equal to it. It results that when the infinite body has completed its revolution, it has traversed an infinite equal to itself in a finite time. But that we know to be impossible.

(7) It can also be shown, conversely, that if the time of revolution is finite, the area traversed must also be finite; but the area traversed was equal to itself; therefore, it is itself finite.

We have now shown that the body which moves in a circle is not endless or infinite, but has its limit.

In another argument involving motion, Aristotle stated that bodies fall with speeds proportional to their weights. This passage may seem hardly relevant to astronomy, but it is historically important. The statement is incorrect; bodies of different weights fall with the same speed. As a weak point in Aristotle's science, his comments on the speeds of falling bodies furnished an opening to critics. The problem of falling motion was an important one in the development of the modern laws of mechanics. The legend of Galileo dropping canon balls from the leaning tower at Pisa, however, may be only a legend.

Excerpt from Aristotle's *De Caelo*

From what we have said, then, it is clear that the weight of the infinite body cannot be finite. It must then be infinite. We have therefore only to show this to be impossible in order to prove an infinite body impossible.

But the impossibility of infinite weight can be shown in the following way. A given weight moves a given distance in a given time; a weight which is as great and more moves the same distance in a less time, the times being in inverse proportion to the weights. For instance, if one weight is twice another, it will take half as long over a given movement. Further, a finite weight traverses any finite distance in a finite time. It necessarily follows from this that infinite weight, if there is such a thing, being, on the one hand, as great and more than as great as the finite, will move accordingly, but being, on the other hand, compelled to move in a time inversely proportionate to its greatness, cannot move at all. The time should be less in proportion as the weight is greater. But there is no proportion between the infinite and the finite: proportion can only hold between a less and a greater finite *time. And though you may say that the time of the movement can be continually diminished, yet there is no minimum. Nor, if there were, would it help us. For some finite body could have been found greater than the given finite in the same proportion which is supposed to hold between the infinite and the given finite; so that an infinite and a finite weight must have traversed an equal distance in equal time. But that is impossible.*

Again, whatever the time, so long as it is finite, in which the infinite performs the motion, a finite weight must necessarily move a certain finite distance in that same time. Infinite weight is therefore impossible, and the same reasoning applies also to infinite lightness. Bodies then of infinite weight and of infinite lightness are equally impossible.

The world was finite, and there was only one world. Aristotle demonstrated this conclusion by yet another *reductio ad absurdum* argument, this one based on the doctrine of natural place. If there were more than one world, each world with a center as the natural place for earthy material to move to and a circumference for fire to move to, then the earth could move toward any of the centers and fire toward any of the circumferences. Chaos would ensue.

Excerpt from Aristotle's *De Caelo*

We must now proceed to explain why there cannot be more than one heaven - the further question mentioned above. For it may be thought that we have not proved universally of bodies that none whatever can exist outside our universe, and that our argument applies only to those of indeterminate extent.

Now all things rest and move naturally and by constraint. A thing moves naturally to a place in which it rests without constraint, and rests naturally in a place to which it moves without constraint. On the other hand, a thing moves by constraint to a place in which it rests by constraint, and rests by constraint in a place to which it moves by constraint. Further, if a given movement is due to constraint, its contrary is natural. If, then, it is by constraint that earth moves from a certain place to the center here, its movement from here to there will be natural, and if earth from there rests here without constraint, its movement hither will be natural. And the natural movement in each case is one.

Further, these worlds, being similar in nature to ours, must all be composed of the same bodies as it. Moreover each of the bodies, fire, I mean, and earth and their intermediates, must have the same power as in our world. For if these names are used equivocally, if the identity of name does not rest upon an identity of form in those elements and ours, then the whole to which they belong can only be called a world by equivocation. Clearly, then, one of the bodies will move naturally away from the center and another towards the center, since fire must be identical with fire, earth with earth, and so on, as the fragments of each are identical in this world. That this must be the case is evident from the principles laid down in our discussion of the movements; for these are limited in number, and the distinction of the elements depends upon the distinction of the movements. Therefore, since the movements are the same, the elements must also be the same everywhere. The particles of earth, then, in another world move naturally also to our center and its fire to our circumference. This, however, is impossible, since if it were true, earth must, in its own world, move upwards, and fire to the center; in the same way the earth of our world must move naturally away from the center when it moves towards the center of another universe. This follows from the supposed juxtaposition of the worlds. For either we must refuse to admit the identical nature of the simple bodies in the various universes, or, admitting this, we must make the center and the extremity one as suggested. This being so, it follows that there cannot be more worlds than one.

By an even more obscure set of arguments, Aristotle demonstrated that the heavens are unalterable. This conclusion would became another weak point in his system, though not until the late sixteenth and early seventeenth centuries A.D. Then, observations of comets moving through the heavens and observations of novae, stars that flare up in brightness, found changes occurring in Aristotle's purportedly unalterable heavens.

Excerpt from Aristotle's *De Caelo*

Having established these distinctions, we may now proceed to the question whether the heaven is ungenerated or generated, indestructible or destructible. . . .

Now to assert that it was generated and yet is eternal is to assert the impossible; for we cannot reasonably attribute to anything any characteristics but those which observation detects in many or all instances. But in this case the facts point the other way: generated things are seen always to be destroyed.

Further, a thing whose present state had no beginning and which could not have been other than it was at any previous moment throughout its entire duration, cannot possibly be changed. For there will have to be some cause of change, and if this had been present earlier it would have made possible another condition of that to which any other condition was impossible. ...

To say that the universe alternately combines and dissolves is no more paradoxical than to make it eternal but varying in shape. It is as if one were to think that there was now destruction and now existence when from a child a man is generated, and from a man a child. For it is clear that when the elements come together the result is not a chance system and combination, but the very same as before - especially on the view of those who hold this theory, since they say that the contrary is the cause of each state. So that if the totality of body, which is a continuum, is now in this order or disposition and now in that, and if the combination of the whole is a world of heaven, then it will not be the world that comes into being and is destroyed, but only its dispositions.

Aristotle went on to show that the heavens rotate and that the earth is stationary in the center. The shape of the heavens is spherical; that is the shape best suited to its nature. The heavens also have a smooth and accurate finish.

Excerpt from Aristotle's *De Caelo*

Since circular motion is not the contrary of the reverse circular motion, we must consider why there is more than one motion, though we have to pursue our inquiries at a distance - a distance created not so much by our spatial position as by the fact that our senses enable us to perceive very few of the attributes of the heavenly bodies. But let not that deter us. The reason must be sought in the following facts.

Everything which has a function exists for its function. The activity of God is immortality, i.e. eternal life. Therefore the movement of that which is divine must be eternal. But such is the heaven, viz. a divine body, and for that reason to it is given the circular body whose nature it is to move always in a circle.

Why, then, is not the whole body of the heaven of the same character as that part? Because there must be something at rest at the center of the revolving body; and of that body no part can be at rest, either elsewhere or at the center. It could do so only if the body's natural movement were towards the center. But the circular movement is natural, since otherwise it could not be eternal: for nothing unnatural is eternal. The unnatural is subsequent to the natural, being a derangement of the natural which occurs in the course of its generation. Earth then has to exist; for it is earth which is at rest at the center. (At present we may take this for granted: it shall be explained later.)

But if earth must exist, so must fire. For, if one of a pair of contraries naturally exists, the other, if it is really contrary, exists also naturally. In some form it must be present, since the matter of contraries is the same. Also, the positive is prior to its privation (warm, for instance, to cold), and rest and heaviness stand for the privation of lightness and movement.

But further, if fire and earth exist, the intermediate bodies must exist also: for each element stands in a contrary relation to every other. (This, again, we will take for granted and try later to explain). . . .

The shape of the heaven is of necessity spherical; for that is the shape most appropriate to its substance and also by nature primary.

First, let us consider generally which shape is primary among planes and solids alike. Every plane figure must be either rectilinear or curvilinear. Now the rectilinear is bounded by more than one line, the curvilinear by one only. But since in any kind the one is naturally prior to the many and the simple to the complex, the circle will be the first of plane figures.

Again, if by complete, as previously defined, we mean a thing outside which no part of itself can be found, and if addition is always possible to the straight line but never to the circular, clearly the line which embraces the circle is complete. If then the complete is prior to the incomplete, it follows on this ground also that the circle is primary among figures.

And the sphere holds the same position among solids. For it alone is embraced by a single surface, while rectilinear solids have several. The sphere is among solids what the circle is among plane figures.

Further, those who divide bodies into planes and generate them out of planes seem to bear witness to the truth of this. Alone among solids they leave the sphere undivided, as not possessing more than one surface: for the division into surfaces is not just dividing a whole by cutting it into its parts, but division of another fashion into parts different in form.

It is clear, then, that the sphere is first of solid figures. . . .

Again, since the whole revolves, palpably and by assumption, in a circle, and since it has been shown that outside the farthest circumference there is neither void nor place, from these grounds also it will follow necessarily that the heaven is spherical. For if it is to be rectilinear in shape, it will follow that there is place and body and void without it. For a rectilinear figure as it revolves never continues in the same room, but where formerly was body, is now none, and where now is none, body will be in a moment because of the projection at the corners. Similarly, if the world had some other figure with unequal radii, if, for instance, it were lentiform, or oviform, in every case we should have to admit space and void outside the moving body, because the whole body would not always occupy the same room. . . .

It is plain from the foregoing that the universe is spherical. It is plain, further, that it is turned (so to speak) with a finish which no manufactured thing nor anything else within the range of our observation can even approach. For the matter of which these are composed does not admit of anything like the same regularity and finish as the substance of the enveloping body; since with each step away from earth the matter manifestly becomes finer in the same proportion as water is finer than earth.

Aristotle next argued that the motion of the heavens is regular. In support of this conclusion he produced many reasons. Again, his arguments are often of the *reductio ad absurdum type*. Aristotle begins with a premise, deduces absurd conclusions, and thus refutes the premise.

Excerpt from Aristotle's *De Caelo*

We have next to show that the movement of the heaven is regular and not irregular. This applies only to the first heaven and the first movement; for the lower spheres exhibit a composition of several movements into one.

If the movement is uneven, clearly there will be acceleration, maximum speed, and retardation, since these appear in all irregular motions. The maximum may occur either at the starting-point or at the goal or between the two; and we expect natural motion to reach its maximum at the goal, unnatural motion at the starting-point, and missiles midway between the two. But circular movement, having no beginning or limit or middle in the direct sense of the words, has neither whence nor whither nor middle: for in time it is eternal, and in length it returns upon itself without a break. If then its movement has no maximum, it can have no irregularity, since irregularity is produced by retardation and acceleration.

Further, since everything that is moved is moved by something, the cause of the irregularity of movement must lie either in the mover or in the moved or in both. For if the mover moved not always with the same force, of if the moved were altered and did not remain the same, of if both were to change, the result might well be an irregular movement in the moved. But none of these possibilities can be conceived as actual in the case of the heavens. As to that which is moved, we have shown that it is primary and simple and ungenerated and indestructible and generally unchanging; and the mover has an even better right to these attributes. It is the primary that moves the primary, the simple the simple, the indestructible and ungenerated that which is indestructible and ungenerated. Since then that which is moved, being a body, is nevertheless unchanging, how should the mover, which is incorporeal, be changed?

It follows then, further, that the motion cannot be irregular. For if irregularity occurs, there must be change either in the movement as a whole, from fast to slow and slow to fast, or in its parts. That there is no irregularity in the parts is obvious, since, if there were, some divergence of the stars would have taken place before now in the infinity of time, as one moved slower and another faster: but no alteration of their intervals is ever observed.

Nor again is a change in the movement as a whole admissible. Retardation is always due to incapacity, and incapacity is unnatural. The incapacities of animals, age, decay, and the like, are all unnatural, due, it seems, to the fact that the whole animal complex is made up of materials which differ in respect of their proper places, and no single part occupies its own place. If therefore that which is primary contains nothing unnatural, being simple and unmixed and in its proper place and having no contrary then it has no place for incapacity, nor, consequently, for retardation or (since acceleration involves retardation) for acceleration.

Again, it is inconceivable that the mover should first show incapacity for an infinite time, and capacity afterwards for another infinity. For clearly nothing which, like incapacity, is unnatural ever continues for an infinity of time; nor does the unnatural endure as long as the natural, or any form of incapacity as long as the capacity. But if the movement is retarded it must necessarily be retarded for an infinite time. Equally impossible is perpetual acceleration or perpetual retardation. For such movement would be infinite and indefinite, but every movement, in our view, proceeds from one point to another and is definite in character.

Again, suppose one assumes a minimum time in less than which the heaven could not complete its movement. For, as a given walk or a given exercise on the harp cannot take any and every time, but every performance has its definite minimum time which is unsurpassable, so, one might suppose, the movement of the heaven could not be completed in any and every time. But in that case perpetual acceleration is impossible (and, equally, perpetual retardation: for the argument holds of both and each), if we may take acceleration to proceed by identical or increasing additions of speed and for an infinite time.

The remaining alternative is to say that the movement exhibits an alternation of slower and faster: but this is a mere fiction and quite inconceivable. Further, irregularity of this kind would be particularly unlikely to pass unobserved, since contrast makes observation easy.

That there is one heaven, then, only, and that it is ungenerated and eternal, and further that its movement is regular, has now been sufficiently explained.

Having sufficiently explained that the motion of the heavens is regular, Aristotle next took up the question of the composition of the stars. This question does not occur in kinematic astronomy, concerned only with describing the motions of the stars. Nor is the question readily susceptible to observational and experimental enquiry until the middle of the nineteenth century A.D. and the advent of the science of spectroscopy. Aristotle argued, persuasively as usual and with inadequate factual basis, that the stars are composed of the same element as the heavens and are fixed to circles which carry them around. His argument is all the more persuasive for the analogy to missiles and heated air; argument by analogy is not to be trusted, in ancient or in modern times.

Excerpt from Aristotle's *De Caelo*

We have next to speak of the stars, as they are called, of their composition, shape, and movements. It would be most natural and consequent upon what has been said that each of the stars should be composed of that substance in which their path lies, since, as we said, there is an element whose natural movement is circular.

In so saying we are only following the same line of thought as those who say that the stars are fiery [a statement with which Aristotle does not agree] *because they believe the upper body to be fire, the presumption being that a thing is composed of the same stuff as that in which it is situated.*

The warmth and light which proceed from them [the stars] are caused by the friction set up in the air by their motion. Movement tends to create fire in wood, stone, and iron; and with even more reason should it have that effect on air, a substance which is closer to fire than these. An example is that of missiles, which as they move are themselves fired so strongly that leaden balls are melted; and if they are fired the surrounding air must be similarly affected. Now while the missiles are heated by reason of their motion in air, which is turned into fire by the agitation produced by their movement, the upper bodies [stars] *are carried on a moving sphere, so that though they are not themselves fired, yet the air underneath the sphere of the revolving body is necessarily heated by its motion, and particularly in that part where the sun is attached to it.*

Hence warmth increases as the sun gets nearer or higher or overhead.

Of the fact, then, that the stars are neither fiery nor move in fire, enough has been said.

Aristotle next asked if the stars are in motion. They do not move of their own effort, but are fixed to spheres and carried about by the spheres.

Excerpt from Aristotle's *De Caelo*

Since changes evidently occur not only in the position of the stars but also in that of the whole heaven, there are three possibilities. Either (1) both are at rest, or (2) both are in motion, or (3) the one is at rest and the other in motion.

(1) That both should be at rest is impossible; for, if the earth is at rest, the hypothesis does not account for the observations; and we take it as granted that the earth is at rest. It remains either that both are moved, or that the one is moved and the other at rest.

(2) On the view, first, that both are in motion, we have the absurdity that the stars and the circles move with the same speed, i.e. that the pace of every star is that of the circle in which it moves. For star and circle are seen to come back to the same place at the same moment; from which it follows that the star has traversed the circle and the circle has completed its own movement, i.e. traversed its own circumference, at one and the same moment. But it is difficult to conceive that the pace of each star should be exactly proportioned to the size of its circle. That the pace of each circle should be proportionate to its size is not absurd but inevitable: but that the same should be true of the movement of the stars contained in the circles is quite incredible. For if, on the one hand, we suppose that the star which moves on the greater circle is necessarily swifter, clearly we also admit that if stars shifted their position so as to exchange circles, the slower would become swifter and the swifter slower. But this would show that their movement was not their own, but due to the circles. If, on the other hand, the arrangement was a chance combination, the coincidence in every case of a greater circle with a swifter movement of the star contained in it is too much to believe. In one or two cases it might not inconceivably fall out so, but to imagine it in every case alike is a mere fiction. Besides, chance has no place in that which is natural, and what happens everywhere and in every case is no matter of chance.

(3) The same absurdity is equally plain if it is supposed that the circles stand still and that it is the stars themselves which move. For it will follow that the outer stars are the swifter, and that the pace of the stars corresponds to the size of their circles.

Since, then, we cannot reasonably suppose either that both are in motion of that the star alone moves, the remaining alternative is that the circles should move, while the stars are at rest and move with the circles to which they are attached. Only on this supposition are we involved in no absurd consequence. For, in the first place, the quicker movement of the larger circle is natural when all the circles are attached to the same center. Whenever bodies are moving with their proper motion, the larger moves quicker. It is the same here with the revolving bodies: for the arc intercepted by two radii will be larger in the larger circle, and hence it is not surprising that the revolution of the larger circle should take the same time as that of the smaller.

And secondly, the fact that the heavens do not break in pieces follows not only from this but also from the proof already given of the continuity of the whole. . . .

That the stars are spherical and are not self-moved, has now been explained.

Finally Aristotle came in his enquiry to the earth. The earth, at the center of the universe, is at rest. Its shape must be spherical; that is the shape it would take as its particles pack into the center. Also, the evidence of the senses indicates that the earth is spherical: eclipses of the moon (when the earth blocks the light from the sun to the moon) reveal that the earth casts a circular shadow. The fact that different stars are seen from different parts of the earth further demonstrates the spherical shape of the earth.

Excerpt from Aristotle's *De Caelo*

It remains to speak of the earth, of its position, of the question whether it is at rest or in motion, and of its shape.

I. As to its position *there is some difference of opinion. Most people - all, in fact, who regard the whole heaven as finite - say it lies at the center. But the Italian philosophers known as Pythagoreans take the contrary view. At the center, they say, is fire, and the earth is one of the stars, creating night and day by its circular motion about the center.*

They further construct another earth in opposition to ours to which they give the name counter-earth. In all this they are not seeking for theories and causes to account for observed facts, but rather forcing their observations and trying to accommodate them to certain theories and opinions of their own.

But there are many others who would agree that it is wrong to give the earth the central position, looking for confirmation rather to theory than to the facts of observation. Their view is that the most precious place befits the most precious thing: but fire, they say, is more precious than earth, and the limit than the intermediate, and the circumference and the center are limits. Reasoning on this basis they take the view that it is not earth that lies at the center of the sphere, but rather fire.

The Pythagoreans have a further reason. They hold that the most important part of the world, which is the center, should be most strictly guarded, and name it, or rather the fire which occupies that place, the "Guard-house of Zeus", as if the word "center" were quite unequivocal, and the center of the mathematical figure were always the same with that of the thing or the natural center. . . .

II. As to the position of the earth, then, this is the view which some advance, and the views advanced concerning its rest or motion *are similar. For here too there is no general agreement. All who deny that the earth lies at the center think that it revolves about the center, and not the earth only but, as we said before, the counter-earth as well. Some of them even consider it possible that there are several bodies so moving, which are invisible to us owing to the interposition of the earth. This, they say, accounts for the fact that eclipses of the moon are more frequent than eclipses of the sun: for in addition to the earth each of these moving bodies can obstruct it. . . .*

III. There are similar disputes about the shape *of the earth. Some think it is spherical, others that it is flat and drum-shaped. . . .*

Others say the earth rests upon water. This, indeed, is the oldest theory that has been preserved, and is attributed to Thales of Miletus. It was supposed to stay still because it floated like wood and other similar substances, which are so constituted as to rest upon water but not upon air. As if the same account had not to be given of the water which carries the earth as of the earth itself! It is not the nature of water, any more than of earth, to stay in mid-air: it must have something to rest upon. . . .

We have now outlined the views held as to the shape, position, and rest or movement of the earth.

Let us first decide the question whether the earth moves or is at rest. For, as we said, there are some who make it one of the stars, and others who, setting it at the center, suppose it to be "rolled" and in motion about the pole as axis.

That both views are untenable will be clear if we take as our starting-point the fact that the earth's motion whether the earth be at the center or away from it, must needs be a constrained motion. It cannot be the movement of the earth itself. If it were, any portion of it would have this movement;

but in fact every part moves in a straight line to the center. Being, then, constrained and unnatural, the movement could not be eternal. But the order of the universe is eternal.

Again, everything that moves with the circular movement, except the first sphere, is observed to be passed, and to move with more than one motion. The earth, then, also, whether it move about the center or is stationary at it, must necessarily move with two motions. But if this were so, there would have to be passings and turnings of the fixed stars. Yet no such thing is observed. The same stars always rise and set in the same parts of the earth.

Further, the natural movement of the earth, part and whole alike, is to the center of the whole - whence the fact that it is now actually situated at the center - but it might be questioned, since both centers are the same, which center it is that portions of earth and other heavy things move to. Is this their goal because it is the center of the earth or because it is the center of the whole? The goal, surely, must be the center of the whole. For fire and other light things move to the extremity of the area which contains the center. It happens, however, that the center of the earth and of the whole is the same. Thus they do move to the center of the earth, but accidentally, in virtue of the fact that the earth's center lies at the center of the whole. That the center of the earth is the goal of their movement is indicated by the fact that heavy bodies moving toward the earth do not move parallel but so as to make equal angles, and thus to a single center, that of the earth.

It is clear, then, that the earth must be at the center and immovable, not only for the reasons already given, but also because heavy bodies forcibly thrown quite straight upward return to the point from which they started, even if they are thrown to an infinite distance. From these considerations then it is clear that the earth does not move and does not lie elsewhere than at the center.

From what we have said the explanation of the earth's immobility is also apparent. If it is the nature of earth, as observation shows, to move from any point to the center, as of fire contrariwise to move from the center to the extremity, it is impossible that any portion of earth should move away from the center except by constraint. For a single thing has a single movement, and a simple thing a simple: contrary movements cannot belong to the same thing, and movement away from the center is the contrary of movement to it. If then no portion of earth can move away from the center, obviously sill less can the earth as a whole so move. For it is the nature of the whole to move to the point to which the part naturally moves. Since, then, it would require a force greater than itself to move it, it must needs stay at the center. . . .

Of the position of the earth and of the manner of its rest or movement, our discussion may here end.

Its shape must necessarily be spherical. For every portion of earth has weight until it reaches the center, and the jostling of parts greater and smaller would bring about not a waved surface, but rather compression and convergence of part and part until the center is reached. The process should be conceived by supposing the earth to come into being in the way that some of the natural philosophers describe. Only they attribute the downward movement to contraint, and it is better to keep to the truth and say that the reason of this motion is that a thing which possesses weight is naturally endowed with a centripetal movement. When the mixture, then, was merely potential, the things that were separated off moved similarly from every side towards the center. Whether the parts which came together at the center were distributed at the extremities evenly, or in some other way, makes no difference. If, on the one hand, there were a similar movement from each quarter of the extremity to the single center, it is obvious that the resulting mass would be similar on every side. For if an equal amount is added on every side the extremity of the mass will be everywhere equidistant from its center, i.e. the figure will be spherical. ...

The evidence of the senses further corroborates this. How else would eclipses of the moon show segments shaped as we see them? As it is, the shapes which the moon itself each month shows are of every kind - straight, gibbous, and concave - but in eclipses the outline is always curved: and, since it is the interposition of the earth that makes the eclipse, the form of this line will be caused by the form of the earth's surface, which is therefore spherical.

Again, our observations of the stars make it evident, not only that the earth is circular, but also that it is a circle of no great size. For quite a small change of position to south or north causes a manifest alteration of the horizon. There is much change, I mean, in the stars which are overhead, and the stars seen are different, as one moves northward or southward. Indeed there are some stars seen in Egypt and in the neighborhood of Cyprus which are not seen in the northerly regions; and stars, which in the north are never beyond the range of observation, in those regions rise and set. All of which goes to show not only that the earth is circular in shape, but also that it is a sphere of no great size: for otherwise the effect of so slight a change of place would not be so quickly apparent.

Hence one should not be too sure of the incredibility of the view of those who conceive that there is continuity between the parts about the pillars of Hercules [peaks at the Strait of Gibraltar, between the Atlantic Ocean and the Mediterranean] *and the parts about India, and that in this way the ocean is one. As further evidence in favor of this they quote the case of elephants, a species occurring in each of these extreme regions, suggesting that the common characteristic of these extremes is explained by their continuity. Also, those mathematicians who try to calculate the size of the earth's circumference arrive at the figure of 400,000 stades* [about 10,000 miles]. *This indicates not only that the earth's mass is spherical in shape, but also that as compared with the stars it is not of great size.*

Aristotle's use of observation, in his arguments that the earth is spherical, is in decided contrast to Plato's rejection of the evidence of the senses. Aristotle's work is thus seen by some historians as a turning point in science, marking the beginning of extensive empirical enquiries.

Even more so in the biological sciences than in the physical sciences, Aristotle made observations of remarkable accuracy. His account of the placental development of the dogfish alone would establish him as a great observer.

Aristotle's methodology is not identical, however, with the methods of modern science. Observation and experiment are employed differently in the development and verification of theory.

Aristotle's scientific method consists of formulating a problem by reviewing the ideas of other philosophers and then seeking out principles from which he can logically infer the conclusions. Observations, cited at the end, are used more to persuade his readers of the truth of his conclusions than as an aid in arriving at the conclusions.

Nor does Aristotle devise critical experiments with which to test his conclusions. And, in comparison to modern scientific methodology, which relies heavily on mathematical quantification, Aristotle's science is largely qualitative.

A satisfactory explanation for Aristotle usually consists of final causes. The teleological and animistic aspect of Aristotle's science is far different from modern mechanical explanations. While bringing to Greek science a new emphasis on the value of observation, Aristotle still differs in important and interesting ways from modern scientific practice.

Aristotle with his concern for causes set a very different problem for scientists than had Plato with his emphasis on forms. The heavens were a topic for scientists following both approaches. Indeed, the same individual might work in both fields. Aristotle, for example, made an important contribution to Plato's problem of saving the phenomena, as will be seen in the next chapter.

A distinction, however, was made, and was formulated in terms of what was the task of the physicist and what was the task of the astronomer. Simplicius in his commentary on Aristotle in the sixth century A.D. described Plato's problem of saving the phenomena; Simplicius also described the difference between the physical and astronomical studies of the heavens.

Excerpt from Simplicius' *Commentary*

To physical theory belongs the study of all that concerns the essence of the heavens and the stars, their power, their quality, their generation and destruction. And, by Zeus, physics also has the power of providing demonstrations concerning the size, shape, and arrangement of these bodies.

Astronomy, on the other hand, is not prepared to say anything about the former. Its demonstrations concern the order of the heavenly bodies, taking it for granted that the heavens are *truly ordered. Astronomy speaks of the shapes, sizes, and relative distance of the earth, the sun, and the moon. It speaks of eclipses, the conjunction of stars, the qualitative and quantitative properties of their movements.*

Now since astronomy depends on the study which considers figures in terms of quality, size, and number, it is quite right that it should require the assistance of arithmetic and geometry. And in dealing with these things, the only ones on which it is authorized to speak, astronomy must conform to arithmetic and geometry.

It happens frequently that the astronomer and the physicist take up the same subject - for instance, they set out to prove that the sun is large or that the earth is round. But in such a case they do not proceed in the same way. The physicist must demonstrate every single one of his propositions by deriving it from the essence of bodies, or from their power, or from what best accords with their perfection, or from their generation and their transformation. The astronomer, on the other hand, establishes his propositions by means of "what goes with" magnitudes and figures, or by means of the magnitude of the motion in question, or the time that corresponds to it. Often the physicist will fasten on the cause and direct his attention to the power that produces the effect he is studying, while the astronomer draws his proofs from circumstances externally related to that same effect. The astronomer is not equipped to contemplate causes, unable to tell us, for instance, what cause is responsible for the spherical shape of the earth and the stars. Sometimes, as for instance when he reasons about eclipses, he does not even try to lay hold of a cause. At other times he feels obliged to posit certain hypothetical modes of being which are such that, once conceded, the phenomena are saved. For example, the astronomer asks why the sun, the moon, and the other wandering stars seem to move irregularly.

Now, whether one assumes that the circles described by the stars are eccentric or that each star is carried along by the revolution of an epicycle, on either supposition the apparent irregularity of their course is saved. The astronomer must therefore maintain that the appearances may be produced by either of these modes of being; consequently his practical study of the movement of the stars will conform to the explanation he has presupposed. This is the reason for Heraclides Ponticus' contention that one can save the apparent irregularity of the motion of the sun by assuming that the sun stays fixed and that the earth moves in a certain way. The knowledge of what is by nature at rest and what properties the things that move have is quite beyond the purviews of the astronomer. He posits, hypothetically, that such and such bodies are immobile, certain others in motion, and then examines with what [additional] suppositions the celestial appearances agree. His principles, namely, that the movements of the stars are regular, uniform, and constant, he receives from the physicist. By means of these principles he then explains the revolutions of all the stars, both those which describe circles parallel to the equator and those which traverse circles that are at an angle to the equator.

The two goals, of Aristotle and of Plato, of physical and of mathematical astronomy, of ascertaining the causes of the motions and of describing the motions, would both be important for the subsequent development of astronomy. Both were set forth in the Hellenic period, between 400 and 300 B.C. In the following Hellenistic period, from 300 B.C. to 150 A.D., Plato's mathematical astronomy received considerable development.

Bibliography

On Aristotle, see the standard account by Werner Jaeger: *Aristotle*. For the treatment of Aristotle as a biologist, see Majorie Grene's *A Portrait of Aristotle*.

Excerpts from Aristotle's *Physica* are from the translation by Richard Hope. Reprinted from ARISTOTLE'S PHYSICS translated by Richard Hope by permission of University of Nebraska Press.

Excerpts from Aristotle's *De Caelo* are from the translation by J. L. Stocks. From *The Oxford Translation of Aristotle*, ed. W. D. Ross, vol. 2 (1930). Reprinted by permission of Oxford University Press.

The excerpt from Simplicius' commentary on Aristotle's *Physica* is from Pierre Duhem's articles on the notion of physical theory from Plato to Galileo, collected and translated by E. Doland and C. Maschler in *To Save the Phenomena*. Reprinted by permission of the University of Chicago Press.

A brief film on this period is *Alexander the Great and the Hellenistic Age* (13 minutes). *Aristotle and the Scientific Method* (14 minutes), in the *Men of Science* series, discusses observation and classification, but at an elementary level.

CHAPTER SIX

Homocentric Spheres

Both Plato and Aristotle in the fourth century B.C. had set problems for astronomers: to save the phenomena of the heavens with a system of uniform circular motions and to understand the cause of the motions. Whether the science of astronomy would make progress toward the solution of either problem depended, in part, upon the level of encouragement and support furnished by society to individual scientists.

Greek society supported playwrights when the citizens of Athens paid to see the productions of Aeschylus, Sophocles, and Euripides, and prizes were offered at festivals for the labors of poets and musicians. Scientists, however, were not as generously supported. Plato lamented that inasmuch as no city held geometry in high regard, inquiries in the subject languished. The practicing physician and architect charged fees for their service. So could the theoretically-inclined scientist or philosopher, were he able to attract a group of pupils. Alternatives were inherited wealth or the beneficence of a wealthy patron.

Science was a fragile enterprise in Greek times, without significant institutional support. It is amazing that the Greeks accomplished as much as they did. What was accomplished owed much to Plato and the haven that his Academy offered for young, would-be philosophers.

To the Academy there came, in about 385 B.C., a poor youth, aged 23, who had travelled to Athens as assistant to a physician. The youth, Eudoxus, stayed only a few months, but sufficiently long to be inculcated with Plato's vision of uniform circular motions. Next the wanderer, with financial assistance from friends in his home town, showed up in Egypt. He stayed there for more than a year, perhaps for several years, and became familiar with the priests' astronomical observations. Eudoxus also made some observations of his own.

Eudoxus established a school and attempted to construct a system of uniform circular motions to produce the observable, apparently irregular motions of the sun, moon, and planets. His system of revolving spheres centered on the earth also satisfied Aristotle's concern for the causes of motion.

Unfortunately, nothing Eudoxus wrote survived. His system is described very briefly by Aristotle in the *Metaphysics*. Nine centuries later another surviving description of Eudoxus' astronomical model was written, a commentary on Aristotle's *On the Heavens* by Simplicius.

Neither description of Eudoxus' work was sufficiently precise to stimulate much interest in the model until the nineteenth century, A.D. Then Giovanni Schiaparelli, an Italian astronomer more widely known for his observation of "canali" on Mars, reconstructed Eudoxus' system.

The reconstruction is compatible with what Aristotle and Simplicius said about Eudoxus. The reconstruction, however, depends on the assumption that Eudoxus' system accounted for many of the astronomical phenomena that we now know. Thus there is danger of attributing to Eudoxus a more advanced knowledge of astronomy than he actually possessed and a more accurate astronomical model. This warning should preface any discussion of what is believed to be Eudoxus' astronomical system.

Excerpt from Aristotle's *Metaphysics*

Now, Eudoxus taught that the paths of the sun and of the moon each encompassed three spheres, the first of these being the sphere of the fixed stars, the second being along the middle line of the zodiac, and the third being inclined across the breadth of the zodiac; the path of the moon is inclined [to the center line of the zodiac] at a greater angle than the path of the sun. And the path of the planets is in each case within four spheres, the first and the second of these also being the same as the first two just mentioned (for the sphere of the fixed stars moves all the spheres, and the sphere placed next inward from this and having its path bisecting the zodiac is common to all), but the third sphere of each planet having its poles in the circle bisecting the zodiac, and the fourth sphere having its path inclined to the equator of the third; the poles of the third sphere being different for each of the other planets, but the poles of Venus and Mercury being the same.

Callippus made the position of the spheres the same as did Eudoxus and assigned the same number as did Eudoxus to Jupiter and to Saturn; but he held that two more spheres are to be added to the sun as well as to the moon, if one is to account for the phenomena, and one more to each of the other planets.

However, if all the spheres combined are to account for the phenomena, there must be for each of the planets other spheres, one less than those enumerated, moving counter to these and bringing back to the same position the outermost sphere of the star [or planet] located inwards from the respective [or further] star [or planet]; for thus alone can all the movements combine to produce the complex movement of the planets.

Summing up, the spheres required by the movements of the planets themselves are: for Saturn and Jupiter, eight; for the others, twenty-five; and of these only those need to have countermoving spheres which are required by the movement of the innermost planet, that is, the spheres moving counter to those of the next two must be six, and the spheres moving counter to those of the next four must be sixteen; so that the total number of moving and countermoving spheres is fifty-five. But if one does not add to the sun and to the moon the movements we have suggested, all the spheres will number only fourty-seven.

A basic observational fact that any astronomical system must account for is the observed movement of the stars overhead each evening. Eudoxus accounted very simply for this phenomenon by placing the stars all on one sphere, which rotated with a uniform speed around the central earth with a period of twenty-four hours. This is equivalent to a rotating earth and a fixed sphere of the stars; the observational consequences of the systems are identical. The relative motion of one star with respect to another had not yet been detected, and thus did not trouble Eudoxus, nor complicate his model.

The apparent motion of the sun presents a slightly more difficult problem. First, there had to be an outer sphere rotating with a period of twenty-four hours, to produce the apparent daily movement of the sun across the sky. Again, as with the stars, the outer sphere for the sun produced the apparent motion now attributed to a rotating earth.

Eudoxus next would have needed a second sphere rotating with a period of a year, to move the sun apparently higher in the sky in summer and lower in winter, and around the heavens with a period of a year. The axis of this second sphere would be tilted relative to the first or outer sphere, to move the sun up in summer and down in winter. This second sphere would have had its axis of rotation fixed to the outer sphere, so it could be carried around with a twenty-four hour period. (The sun is seen high in the summer sky and low in the winter sky because the earth's axis of rotation is not perpendicular to the plane of the earth's orbit around the sun, but is inclined at an angle of about 23.5 degrees. Thus the sun's apparent annual motion carries it alternately above and below the plane of the earth's equator.)

The Sun high in Summer Sky, low in Winter Sky

A. Earth orbiting the sun.

Winter in the Northern Hemisphere

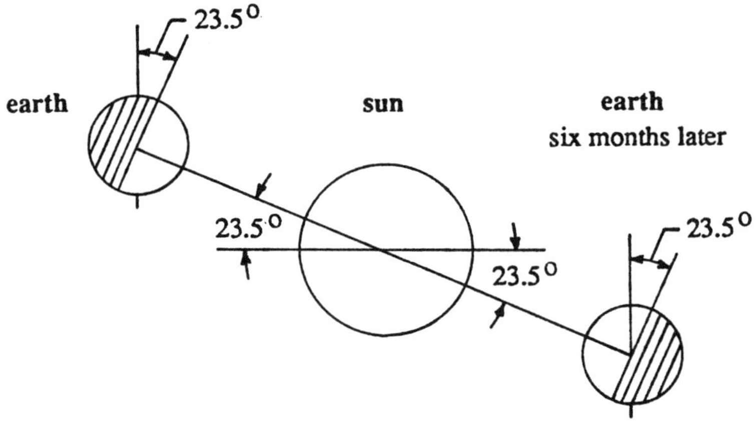

Summer in the Northern Hemisphere

B. Position of the sun as seen from the earth.

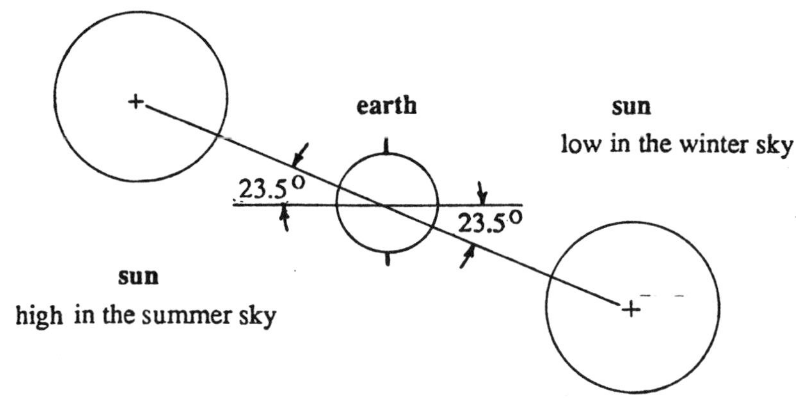

Eudoxus ignored the changeable velocity of the sun, already discovered by his time. This decision, whatever its justification, saved him much trouble. Eudoxus did add a third sphere, though, to account for an observation now known to have been mistaken. Continually it would be a problem to decide which observations were accurate and should be incorporated into a model. A rather uncritical acceptance of observational reports produced difficulties that might otherwise have been avoided.

Eudoxus devised a similar system of spheres for the moon. First, as with all celestial objects, there was the outer sphere rotating once every twenty-four hours. This outer sphere produced the phenomena now known to be produced by the daily rotation of the earth.

The moon circles the earth approximately once a month. To produce the monthly motion, Eudoxus added a second sphere attached to the first and rotating west to east with a period of one lunar month.

There is also a small variation in the latitude of the moon. The moon moves at times slightly above and at other times slightly below an imaginary plane containing the earth and the sun. The modern explanation for this observed phenomenon is that the plane of the moon's orbit around the earth is inclined at an angle of approximately five degrees to the plane of the earth's orbit around the sun. Eudoxus added a third sphere to produce the small deviations in the latitude of the moon.

Planetary motions presented a more difficult problem than did the motions of either the sun or the moon. The planets display a retrograde motion; they appear at times to cease their motion relative to the stars, turn back temporarily, retrace a small part of their path, and then change direction once more and resume their voyages around the heavens. Eudoxus' task was to devise a model consisting of uniform circular motions only yet producing the apparent retrograde motions observed.

Eudoxus first gave each planet an outer sphere to carry it around the earth with a period of twenty-four hours. Second spheres moved the planets around the heavens, with periods of a year for Venus and for Mercury and with longer periods for the outer planets. To produce observed motions in latitude, Eudoxus added third spheres for each planet. So far, the planetary solutions followed the solutions for the sun and the moon.

To produce the observed retrograde motions, Eudoxus added a fourth sphere. By a clever combination of inclinations and speeds of revolution of the third and fourth spheres, Eudoxus was able to produce in an approximate fashion the observed retrograde motions.

Retrograde motion can be produced from a combination of two spheres rotating with unchanging velocities. The following demonstration may suggest an approximation of Eudoxus' system, whose details have been

Apparent Retrograde Motion of a Planet

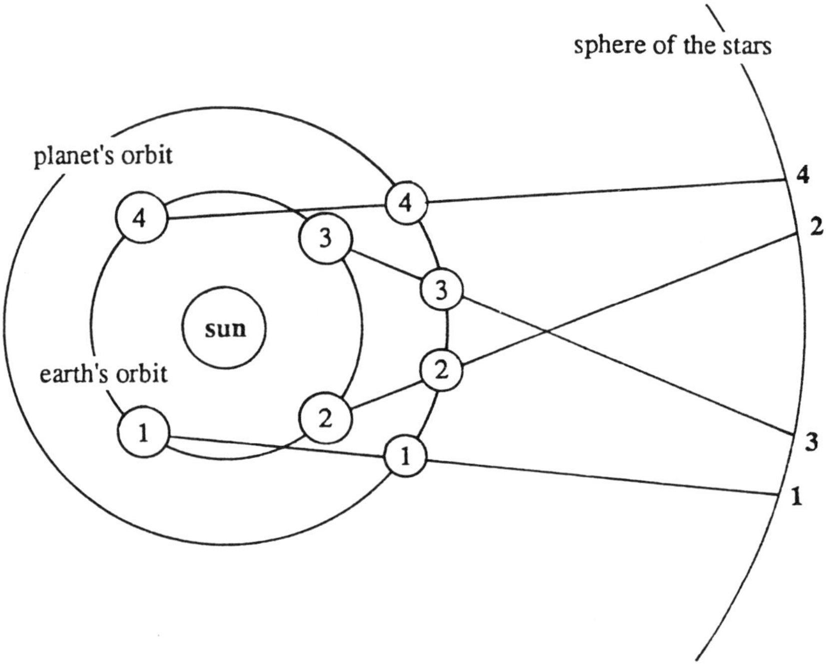

As seen from earth at times 1, 2, 3, and 4, the planet apparently moves against the sphere of the stars from 1 to 2, turns back to 3, and then resumes its forward motion to 4. The earth, moving faster than the outer planet, overtakes and passes it.

lost to time. The two spheres have the same center; they are homocentric. The outer sphere is not shown in the drawing. Its axis of rotation is vertical, in the plane of the paper, in a north-south direction. The outer sphere carries everything within it eastward. The axis of the inner sphere is horizontal and in the plane of the paper. The motion of a planet carried about by the inner sphere is up (north) and down (south) and into (west) and out of (east) the plane of the paper. The planet appears to move north and west from 1 to 2, north and east to 3, south and east to 4, and south and west back to 1. When the inner sphere imparts an eastward motion to the planet, moving it from 2 to 3 to 4, the total eastward motion, including the steady eastward motion imparted by the outer sphere, will be very rapid. If the westward speed imparted by the inner sphere is greater than the steady eastward motion imparted by the outer sphere, then the planet will appear to slow down and briefly move west during the passage from 4 to 1 to 2. Thus retrograde motion can be produced from a combination of two spheres rotating with constant velocities.

Retrograde Motion from Homocentric Spheres

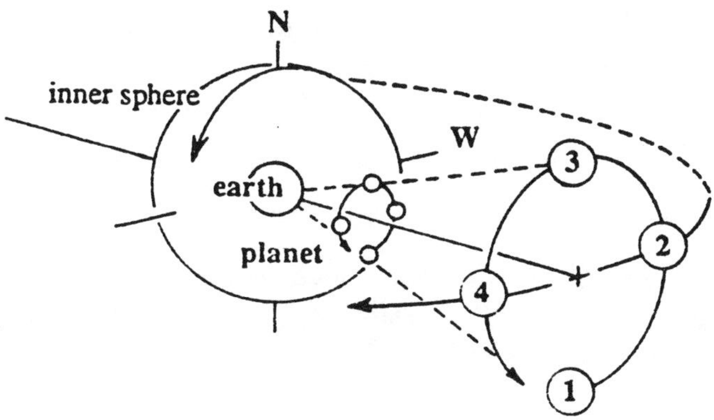

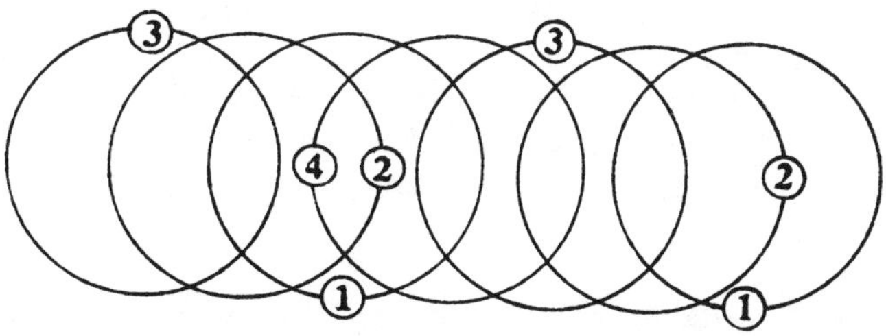

The apparent path of the planet as seen against the sphere of the stars due to the motions of the inner and outer spheres.

A system of four homocentric spheres, however, cannot simultaneously produce with complete quantitative accuracy both the length of the retrograde motion westward and the length of the motion in latitude (north-south) for all the planets. Eudoxus' contemporaries apparently detected flaws in his system, because there occurred a series of modifications after his death in 355 B.C.

Work on Eudoxus' spherical system was continued, first at his school in Cyzicus, a city on the south coast of the Sea of Marmara (now Turkey), by his pupil, Polemarchus, and then by Polemarchus' pupil, Callippus. Later, the work was continued in Athens, where both Polemarchus and Callippus moved. There Callippus also worked with Aristotle.

According to Aristotle, in his *Metaphysics*, Callippus was satisfied with Eudoxus' system for Jupiter and Saturn, but added additional spheres for the other planets, for the sun, and for the moon. Aristotle, not content with a purely geometrical system but also concerned with how the combination of forces caused the motions of the planets, found it necessary to add several more spheres to the system.

Simplicius' *Commentary* on *On the Heavens*

Nevertheless the theories of Eudoxus and his followers fail to save the phenomena, and not only those which were first noticed at a later date, but even those which were before known and actually accepted by the authors themselves.

What need is there for me to mention the generality of these, some of which, after Eudoxus had failed to account for them, Callippus tried to save, - if indeed we can regard his as so far successful? I confine myself to one fact which is actually evident to the eye; this fact no one before Autolycus of Pitane even tried to explain by means of hypotheses, and not even Autolycus was able to do so, as clearly appears from his controversy with Aristotherus.

I refer to the fact that the planets appear at times to be near to us and at times to have receded. This is indeed obvious to our eyes in the case of some of them; for the star called after Aphrodite [Venus] and also the star of Ares [Mars] seem, in the middle of their retrogradations, to be many times as large, so much so that the star of Aphrodite actually makes bodies cast shadows on moonless nights.

The moon also, even in the perception of our eye, is clearly not always at the same distance from us, because it does not always seem to be the same size under the same conditions as to medium. The same fact is moreover confirmed if we observe the moon by means of an instrument; for it is at one time a disc of eleven fingerbreadths, and again at another time a disc of twelve fingerbreadths, which when placed at the same distance from the observer hides the moon (exactly) so that his eye does not see it.

In addition to this, there is evidence for the truth of what I have stated in the observed facts with regard to total eclipses of the sun; for when the center of the sun, the center of the moon, and our eye happen to be in a straight line, what is seen is not always alike; but at one time the cone which comprehends the moon and has its vertex at our eye comprehends the sun itself at the same time, and the sun even remains invisible to us for a certain time, while again at another time this is so far from being the case that a rim of a certain breadth on the outside edge is left visible all round it at the middle of the duration of the eclipse. Hence we must conclude that the apparent difference in the sizes of the two bodies observed under the same atmospheric conditions is due to the inequality of their distances (at different times) ....

But indeed this inequality in the distances of each star at different times cannot even be said to have been unknown to the authors of the concentric theory themselves. For Polemarchus of Cyzicus appears to be aware of it, but to minimize it as being imperceptible, because he preferred the theory which placed the spheres themselves about the very center in the universe. Aristotle, too, shows that he is conscious of it when, in the Physical Problems, *he discusses objections to the hypotheses of astronomers arising from the fact that even the sizes of the planets do not appear to be the same always. In this respect Aristotle was not altogether satisfied with the revolving spheres, although the supposition that, being concentric with the universe, they move about its center attracted him. Again, it is clear from what he says in Book Λ of the* Metaphysics *that he thought that the facts about the movements of the planets had not been sufficiently explained by the astronomers who came before him or were contemporary with him. At all events we find him using language of this sort: "(on the question how many in number these movements of the planets are), we must for the present content ourselves with repeating what some of the mathematicians say, in order that we may form a notion and our mind may have a certain definite number to apprehend; but for the rest we must investigate some matters for ourselves and learn others from other investigators, and, if those who study these questions reach conclusions different from the views now put forward, we must, while respecting both, give our adherence to those which are the more correct."*

Eudoxus' astronomical system is important in the history of science and in the history of civilization. His scientific model was one of the first major attempts to go beyond solely philosophical reasoning and to develop a model in considerable detail and in agreement with observed phenomena.

The development of Eudoxus' system by Callippus and Aristotle, and possibly by others after them, is also important. The subsequent development furnishes an early example of continuity in science and the cummulative advance possible when a problem receives continued attention over the lifetimes of several generations of scientists.

Such scientific progress depends upon society. Persistent scientific advance, as distinct from scientific revolutions, is encouraged by some stability in working conditions and the means to transmit knowledge from one generation to the next. Science is an organized social activity conditioned by the society in which it exists.

While Eudoxus' system was improved and brought into better agreement with observed planetary motions, there was one phenomenon for which it could not account. As Simplicius wrote in his commentary on Aristotle's *On the Heavens*, the planets move at different times closer to or farther from the earth. In the case of the moon, apparent changes in size (due to changing distance from the observer on the earth) were observed directly by the Greeks. For the planets, apparent changes in size and hence in actual distance could be inferred from changes in apparent brightness. (According to Aristotle, there was no change in the heavens; thus an apparent change in brightness was not real, and was to be explained by a change in distance from the observer.)

A system of homocentric spheres cannot produce changes in the distance of objects from the center of the spheres, where the earth was thought to reside. While Eudoxus may have been solely concerned with the movements in longitude and latitude, the problem of the distances seems to have been accepted as a legitimate problem by Greek astronomers working immediately after Eudoxus.

Seemingly no one proposed the simple expedient of centering the spheres about a point near but not exactly coincident with the earth. Or if anyone did, he quickly rejected the idea without much further thought, because Aristotelian physics required that the earth be in its natural place at the center of the universe. Also, there may well have been aesthetic objections, articulated or not, to off-center celestial spheres.

Whatever the reason, Eudoxus' system disappeared from history. It was not recovered until more than two thousand years later. Now it is an important historical artifact, rather than a plausible scientific model of the universe.

Bibliography

For a review of planetary motions and the system of homocentric spheres, see Thomas Kuhn's *The Copernican Revolution*, especially chapter two on "The Problem of the Planets".

The classic work on the history of planetary astronomy is by J. L. E. Dreyer, *History of the Planetary System from Thales to Kepler*, reprinted as *A History of Astronomy from Thales to Kepler*.

On Eudoxus, see Erkka Maula, *Studies in Eudoxus' Homocentric Spheres*.

The excerpt from Aristotle's *Metaphysics* is from the translation by Richard Hope. It is reprinted here by permission of Columbia University Press.

The excerpt from Simplicius' Commentary on Aristotle's *On the Heavens* is from *Aristarchus of Samos: The Ancient Copernicus* trans. Sir Thomas Heath (1913). Reprinted by permission of Oxford University Press.

A film on the general historical problem of the planetary motions from the Babylonians to 1700 is *Earth and Sky* (45 minutes).

Two short film loops showing retrograde motion from different perspectives are *Retrograde Motion: Heliocentric Model* (4 minutes) and *Retrograde Motion: Geocentric Model* (4 minutes).

The problem of changing planetary distance is apparent in the above films and in the short film loop *Planetary Motion*.

CHAPTER SEVEN

Of Geometry and Men

One alternative to Eudoxus' system of homocentric spheres was suggested by Aristarchus. He was a pupil of Strato, who succeeded Theophrastus as the head of Aristotle's school, the Lyceum, in about 287 B.C. Aristarchus may have studied with Strato at Athens, or earlier in Alexandria, Egypt, before Strato became head of the Lyceum.

It is with Aristarchus, either just before him or just after him, that the center of scientific activity shifted from Athens to Alexandria. And it is by Aristarchus that some of the most beautiful of Greek geometry was created.

According to Archimedes, a younger contemporary of Aristarchus and a student in Alexandria before returning to his native Syracuse, Aristarchus proposed a system with the sun in the center, the earth orbiting the sun, and the outer sphere of the stars motionless.

Excerpt from Archimedes' *Sand-reckoner*

You are aware that "universe" is the name given by most astronomers to the sphere, the center of which is the center of the earth, while its radius is equal to the straight line between the center of the sun and the center of the earth. This is the common account, as you have heard from astronomers.

But Aristarchus brought out a book consisting of certain hypotheses, *wherein it appears, as a consequence of the assumptions made, that the universe is many times greater than the "universe" just mentioned. His hypotheses are that* the fixed stars and the sun remain unmoved, that the earth revolves about the sun in the circumference of a circle, the sun lying in the middle of the orbit, *and that the sphere of the fixed stars, situated about the same center as the sun, is so great that the circle in which he supposes the earth to revolve bears such a proportion to the distance of the fixed stars as the center of the sphere bears to its surface.*

Now it is easy to see that this is impossible; for, since the center of the sphere has no magnitude, we cannot conceive it to bear any ratio whatever to the surface of the sphere. We must, however, take Aristarchus to mean this: since we conceive the earth to be, as it were, the center of the universe, the ratio which the earth bears to what we describe as the "universe" is equal to the ratio which the sphere containing the circle in which he supposes the earth to revolve bears to the sphere of the fixed stars. For he adapts the proofs of the phenomena to a hypothesis of this kind, and in particular he appears to suppose the size of the sphere in which he makes the earth move to be equal to what we call the "universe".

Aristarchus' hypothesis is a remarkable anticipation of the Copernican theory, to be established eighteen-hundred years later. Why did Greek philosophers show virtually no interest in Aristarchus' hypothesis? There are several possible answers, any combination of which might plausibly account for the lack of interest.

First, a model of the universe with the earth revolving around the sun is not compatible with Aristotleian physics. In Aristotle's physics, the natural place of earthy material is at the center of the universe, and the laws of motion depend upon that arrangement.

Second, there may have been religious objections to moving the sacred earth from the center. Indeed, according to a later text, someone did suggest that a charge of impiety be brought against Aristarchus. There is no evidence, however, that anyone acted on the suggestion.

Third, there would have been observational problems with Aristarchus' model, the same that later critics would raise against Copernicus' heliocentric theory, in the sixteenth century A.D. If the earth does orbit the sun, the apparent positions of stars as measured from the earth (their parallax angles) change as the earth moves. A simple illustration of parallax is created when a person determines the angular direction of an object in front of him or her, then walks past it, and finds that the position of the object relative to the moving person (or to a stationary person moved by a moving earth) has changed.

Unable to detect a parallax of the stars, Copernicus would have to resort to the argument that the stars are so far distant from the earth that the parallax angle, consequently minute, is too small to measure. Copernicus' explanation is correct; in 1838 a stellar parallax finally was measured, with vastly improved instruments. However, the argument that something exists but is too small to detect may well be reasonably viewed with some skepticism. Perhaps the same objection to Copernicus' theory

Stellar Parallax

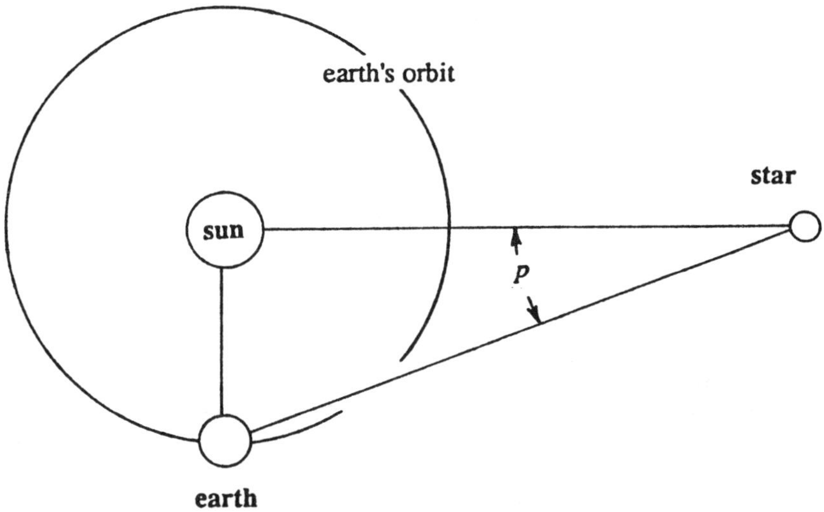

The parallax angle *p* decreases as the distance of the star from the earth increases. Often it is too small to detect.

was raised against Aristarchus. He did write that the sphere of the fixed stars is very large relative to the size of the earth's orbit. The parallax angle p is smaller, the more distant the star, as can be readily seen in the above drawing illustrating stellar parallax.

Yet a fourth possible explanation for the neglect of Aristarchus' heliocentric system is that it wasn't a system - that Aristarchus never developed his hypothesis in sufficient detail to show that it could account for the observed phenomena. Philosophers were turning away from unsupported theorizing and placing a new emphasis on observation.

Fifth, in the shift of work in astronomy from Athens to Alexandria then underway, Aristarchus' idea may just have been lost sight of, if not by Archimedes, at least by most other philosophers.

Aristarchus' heliocentric theory may have been presented only vaguely, but his calculation of the sizes and distances of the sun and moon was presented in exquisite geometrical detail. Surprisingly, in view of his heliocentric hypothesis, Aristarchus assumed a geocentric model in his distance calculations, with the sun and moon moving in circles around the earth at the center.

Historians have suggested that the work on sizes and distances was done before Aristarchus began to think about a heliocentric model. The assumption is that once Aristarchus had hit upon the true heliocentric model, he would not have reverted to an earlier and incorrect model of the universe. Other explanations, though, are possible. Aristarchus did not know with certainty which model, heliocentric or geocentric, was a more true representation of the universe. Nor is it impossible that he abandoned the heliocentric model and later wrote his treatise on sizes and distances. Also, in using a geocentric model for his calculations of sizes and distances, Aristarchus may have chosen the model most familiar to others and thus the more likely to be accepted. For the limited purpose of calculating the ratio of the distance of the sun and moon from the earth, it matters not whether a geocentric or heliocentric model is employed.

That Aristarchus was more concerned with the geometrical principles involved than with an accurate picture of the universe becomes apparent when he introduces the angular diameter of the sun into his calculation. He had accurately measured the size, according to Archimedes, finding a value of about half a degree for the angular diameter of the sun. Possibly Aristarchus used an improved sun dial to make his observations; he is credited with improving that instrument. In his treatise on sizes and distances, however, Aristarchus uses a value of two degrees for the angular diameter of the sun. It may be that Aristarchus meant to write, or originally wrote, "one-fiftieth part of a sign of the zodiac", that is, one-fiftieth of thirty degrees, which is close to the actual observed value; and the word was inadvertently changed in the process of copying or translating to one-fifteenth, that is, one-fifteenth of thirty degrees, or two degrees. Aristarchus could, though, have meant to use the value of two degrees: possibly to emphasize the principle of the problem rather than the answer, possibly to avoid a controversy over what was the observed value, and also possibly to simplfy the sample calculations.

The calculation begins with a list of hypotheses: the moon receives its light from the sun; the moon subtends one-fifteenth part of a sign of the zodiac, i.e. one-fifteenth of 30, or 2 degrees; when the moon appears halved the great circle dividing the dark and light parts of the moon is in the direction of the observer's eye; and when the moon appears halved it is distant from the sun less than a quadrant, i.e. 90 degrees, by one-thirtieth of a quadrant, i.e. 3 degrees, or is 87 degrees (ninety minus three) distant from the sun.

From this data, which is far from the truth, and hence the answer will also be similarly far from the truth, Aristarchus is able to show, after laying some mathematical groundwork, that the sun's distance from the earth would be between 18 and 20 times the moon's distance from the earth.

The geometrical argument has a spare economy and elegance, not to be improved upon by later commentaries or précis.

Aristarchus' *Sizes and Distances of Sun and Moon*

Hypotheses

1. That the moon receives its light from the sun.
2. That the earth is in the relation of a point and center to the sphere in which the moon moves.
3. That, when the moon appears to us halved, the great circle which divides the dark and the bright portions of the moon is in the direction of our eye.
4. That, when the moon appears to us halved, its distance from the sun is then less than a quadrant by one-thirtieth of a quadrant.
5. That the breadth of the [earth's] shadow is [that] of two moons.
6. That the moon subtends one-fifteenth part of a sign of the zodiac.

We are now in a position to prove the following propositions:

1. The distance of the sun from the earth is greater than eighteen times, but less than twenty times, the distance of the moon (from the earth); this follows from the hypothesis about the halved moon.
2. The diameter of the sun has the same ratio (as aforesaid) to the diameter of the moon.
3. The diameter of the sun has to the diameter of the earth a ratio greater than that which 19 has to 3, but less than that which 43 has to 6; *this follows from the ratio thus discovered between the distances, the hypothesis about the shadow and the hypothesis that the moon subtends one-fifteenth part of a sign of the zodiac.*

Proposition 7

The distance of the sun from the earth is greater than eighteen times, but less than twenty times, the distance of the moon from the earth.

For let A *be the center of the sun,* B *that of the earth.*

Let AB *be joined and produced.*

Let C *be the center of the moon when halved; let a plane be carried through* AB *and* C, *and let the section made by it in the sphere on which the center of the sun moves be the great circle* ADE.

Let AC, CB *be joined, and let* BC *be produced to* D.

Then, because the point C *is the center of the moon when halved, the angle* ACB *will be right.*

Let BE *be drawn from* B *at right angles to* BA; *then the circumference* ED *will be one-thirtieth of the circumference* EDA; *for, by hypothesis, when the moon appears to us halved, its distance from the sun is less than a quadrant by one-thirtieth of a quadrant.*

Thus the angle EBC *is also one-thirtieth of a right angle.*

Let the parallelogram AE *be completed, and let* BF *be joined.*

Then the angle FBE *will be half a right angle.*

Let the angle FBE *be bisected by the straight line* BG; *therefore the angle* GBE *is one-fourth part of a right angle.*

But the angle DBE *is also one-thirtieth part of a right angle; therefore the ratio of the angle* GBE *to the angle* DBE *is that which 15 has to 2; for, if a right angle be regarded as divided into 60 equal parts, the angle* GBE *contains 15 of such parts, and the angle* DBE *contains* 2.

Now, since GE *has to* EH *a ratio greater than that which the angle* GBE *has to angle* DBE, *therefore* GE *has to* EH *a ratio greater than that which 15 has to 2.*

Next, since BE *is equal to* EF, *and the angle* E *is right, therefore the square on* FB *is double of the square on* BE.

But, as the square on FB *is to the square on* BE, *so is the square on* FG *to the square on* GE; *therefore the square on* FG *is double of the square on* GE.

Now 49 is less than double of 25, so that the square on FG *has to the square on* GE *a ratio greater than that which 49 has to 25; therefore* FG *also has to* GE *a ratio greater than that which 7 has to 5.*

Therefore, componendo, FE *has to* EG *a ratio greater than that which 12 has to 5, that is, than that which 36 has to 15.*

But it was also proved that GE *has to* EH *a ratio greater than that which 15 has to 2; therefore,* ex aequali, FE *has to* EH *a ratio greater than that which 36 has to 2, that is, than that which 18 has to 1; therefore* FE *is greater than 18 times* EH.

And FE *is equal to* BE; *therefore* BE *is also greater than 18 times* EH; *therefore* BH *is much greater than 18 times* HE.

But, as BH *is to* HE, *so is* AB *to* BC, *because of the similarity of the triangles; therefore* AB *is also greater than 18 times* BC.

And AB *is the distance of the sun from the earth, while* CB *is the distance of the moon from the earth; therefore the distance of the sun from the earth is greater than 18 times the distance of the moon from the earth.*

Again, I say that it is also less than 20 times that distance.

For let DK *be drawn through* D *parallel to* EB, *and about the triangle* DKB *let the circle* DKB *be described; then* DB *will be its diameter, because the angle at* K *is right.*

Let BL, *the side of a hexagon, be fitted into the circle.*

Then, since the angle DBE *is one-thirtieth of a right angle, the angle* BDK *is also one-thirtieth of a right angle; therefore the circumference* BK *is one-sixtieth of the whole circle.*

But BL *is also one-sixth part of the whole circle.*

Therefore the circumference BL *is ten times the circumference* BK.

And the circumference BL *has to the circumference* BK *a ratio greater than that which the straight line* BL *has to the straight line* BK; *therefore the straight line* BL *is less than ten times the straight line* BK.

And BD *is double of* BL; *therefore* BD *is less than 20 times* BK.

But, as BD *is to* BK, *so is* AB *to* BC; *therefore* AB *is also less than 20 times* BC.

And AB *is the distance of the sun from the earth; while* BC *is the distance of the moon from the earth; therefore the distance of the sun from the earth is less than 20 times the distance of the moon from the earth.*

And it was before proved that it is greater than 18 times that distance.

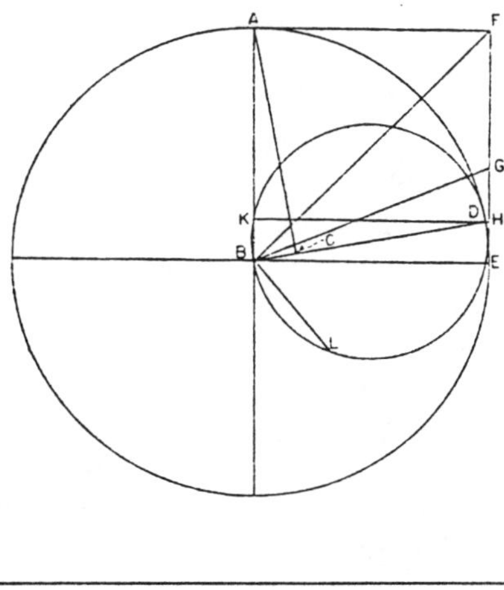

Since the sun and the moon have almost exactly the same angular diameter as measured from the earth, their sizes are in the same ratio as their distances: the sun is more than 18 times larger than the moon but less than 20 times larger.

Cone of the Sun and Moon observed from Earth

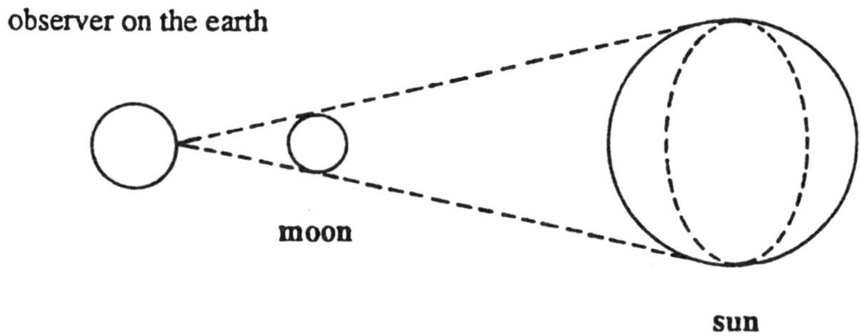

Aristarchus continued his bravado display of geometrical wizardry, proving that the diameter of the moon is less than 2/45 but greater than 1/31 of its distance to the earth and that the diameter of the sun has to the diameter of the earth a ratio greater than that which 19 has to 3 but less than that which 43 has to 5. Aristarchus also calculated ratios for the volumes of the earth, moon, and sun, knowing as he did that the volume is as the cube of the diameter.

Aristarchus delighted in the sheer beauty of geometry. It mattered little to him whether the ratios were accurate. There was scarcely any practical use for a knowledge of the ratios of distances in the solar system. What Aristarchus' investigation achieved was an elevation of souls, the soul of the investigator and the souls of those who study his geometry. That, some might argue, is the most practical achievement of all.

Beautiful as it is, Aristarchus' geometrical creation falls outside the major astronomical tradition of saving the phenomena with a system of uniform circular motions. His friend Eratosthenes' measurement of the circumference of the earth is in the same category, outside the astronomical tradition and a beautiful geometrical work of art and intellect.

Eratosthenes, who lived from about 276 to 195 B.C., studied in Athens but did most of his work in Alexandria. It was there and at Syene, to the south, that Eratosthenes made his famous measurement of the size of the earth.

Eratosthenes noticed that at mid-day in the middle of summer no shadow was cast by a vertical stick (actually, the pointer of a sundial) at Syene and that the sun's rays reached to the bottom of a well. Thus the sun was directly overhead at that instant. Almost due north at Alexandria, on the same day at the same time, the sun was observed to cast a shadow.

The angle of the shadow revealed the size of the earth. Since the rays of the sun reaching the earth can be treated as parallel, because of the enormous distance from the sun to the earth, and vertical sticks point to the center of the earth, the angle of the shadow at Alexandria is the same as the angle between radius vectors drawn from the center of the earth to Alexandria and to Syene.

Eratosthenes and the Size of the Earth

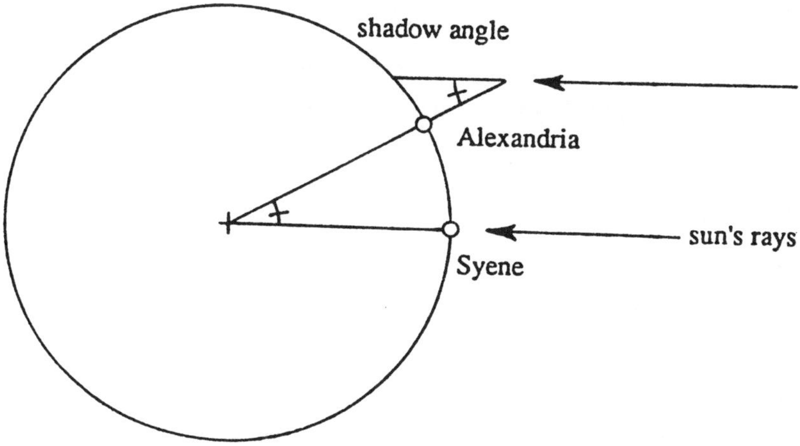

This angle Eratosthenes found to be 1/50 of a circle, slightly more than seven degrees. The distance between Syene and Alexandria was approximately 5,000 stades. Therefore the circumference of the earth, fifty times greater than the distance between Alexandria and Syene, was calculated to be 250,000 stades.

Not knowing how to translate stades into modern units, it is a bit difficult to assess the accuracy of Eratosthenes' measurement of the size of the earth. His result would have been slightly in error because Alexandria is not quite due north of Syene; it lies three degrees west. Furthermore, Eratosthenes' estimate of the distance between Alexandria and Syene may not have been exact. Possibly Eratosthenes' value was equal to about 29,000 miles, not terribly far above the modern value of just under 25,000 miles for the circumference of the earth. Even were his value badly wrong, the method was correct in principle.

Estimates of the size of the earth were to play a part in the voyage of Cristopher Columbus in 1492 A.D., though it was not Eratosthenes' reasonably large value that Columbus chose to emphasize in his attempts to persuade the Spanish authorities to pay for a venture nearly as costly as it was bold. Nor was it Aristotle's considerably larger estimate, also known in the middle ages, that Columbus chose to use. Aristotle, in *On the Heavens*, had noted that different stars are seen overhead from different places on the earth, showing that the earth is circular in shape and of no great size relative to the sphere of the stars, else the effect of a slight change in place would not be so quickly apparent. Stating neither the method used nor those using it, Aristotle wrote that mathematicians had calculated the earth's circumference to be 400,000 stades.

Excerpt from Aristotle's *On the Heavens*

Again, our observations of the stars make it evident, not only that the earth is circular, but also that it is a circle of no great size.

For quite a small change of position to south or north causes a manifest alteration of the horizon. There is much change, I mean, in the stars which are overhead, and the stars seen are different, as one moves northward or southward. Indeed there are some stars seen in Egypt and in the neighborhood of Cyprus which are not seen in the northerly regions; and stars, which in the north are never beyond the range of observation, in those regions rise and set. All of which goes to show not only that the earth is circular in shape, but also that it is a sphere of no great size: for otherwise the effect of so slight a change of place would not be so quickly apparent.

Hence one should not be too sure of the incredibility of the view of those who conceive that there is continuity between the parts about the pillars of Hercules [peaks on both sides of the Straight of Gibraltar, betweeen the Atlantic ocean and the Mediterranean] and the parts about India, and that in this way the ocean is one.

As further evidence in favor of this they quote the case of elephants, a species occurring in each of these extreme regions, suggesting that the common characteristic of these extremes is explained by their continuity. Also, those mathematicians who try to calculate the size of the earth's circumference arrive at the figure 400,000 stades. This indicates not only that the earth's mass is spherical in shape, but also that as compared with the stars it is not of great size.

Aristotle's exaggerated estimate of the size of the earth would not have offered much encouragement to prospective backers of Columbus' venture westward in hope of reaching the spice islands, now known as the East Indies, as opposed to the West Indies, which Columbus accidentally discovered well short of his intended goal. Another value for the size of the earth was available in the Middle Ages, though, calculated by Arab scientists using the Greek geometrical method of Posidonius. It was this smaller and more propitious figure that Columbus chose.

A variation on the method of Eratosthenes for determining the size of the earth was devised by Posidonius, a Greek philosopher who lived from approximately 135 to 51 B.C. He studied at Athens, later established himself as a teacher at Rhodes, and served as an ambassador to Rome. Distinguished Romans, including Cicero, studied with him.

Posidonius wrote many books on astronomy, geography, and history, all of them lost. A description of his' method was preserved by Cleomedes, a later Greek astronomer, in his book *On the Orbits of the Heavenly Bodies.*

Posidonius noted that Rhodes and Alexandria lie on the same meridian circle, drawn through the poles of the universe. In other words, the cities are at the same longitude on the earth; Alexandria is due south of Rhodes. It was the same type of relationship as that between Alexandria and Syene exploited earlier by Eratosthenes.

At Rhodes, the bright star Canopus was seen very nearly on the horizon. Farther north, the star was not visible. Some 5,000 stades farther south, at Alexandria, the star was observed at a height above the horizon of one-fourth of a sign of the zodiac, or one fourty-eighth part of a circle, that is, seven and a half degrees.

The light rays from the star are approximately parallel, following the same argument Eratosthenes used for the rays from the sun. Therefore the distance between Rhodes and Alexandria is one forty-eighth of the circumference of the earth; or, the circumference of the earth was 240,000 stades. Such was Posidonius' way of dealing with the size of the earth.

Posidonius and the Size of the Earth

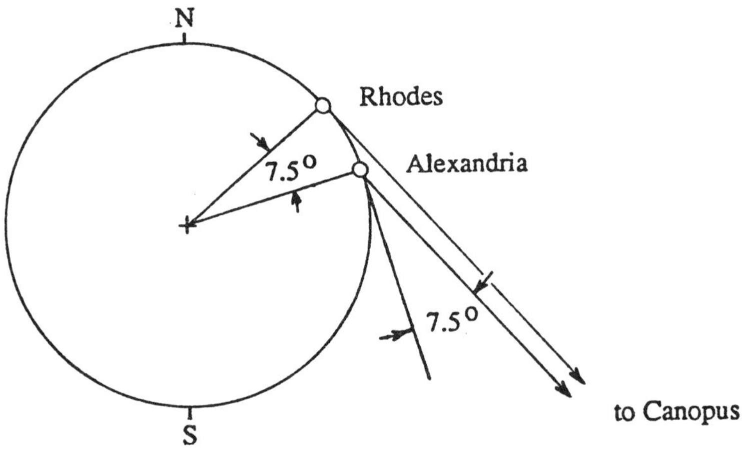

Geometry and astronomy did play a role in Columbus' voyages, not only in estimates of the size of the earth and consequently the estimated duration of the trip, but also in extricating Columbus and his crew from a tight spot. To persuade recalcitrant Indians (Columbus thought he had arrived in the East Indies, not in a new world) that God was angry with their failure to keep Columbus and his men amply supplied with food, Columbus warned the Indians that that very night they would see a token of the punishment that God was going to visit upon them. The Indians' merriment turned to concern as the moon rose, inflammed with wrath. They hurried off to find supplies. Columbus retired to his cabin, not to pray to God to forgive the Indians, as he had assured them he would, but to calculate from his observation of the lunar eclipse the longitude of his ship. He had calculated longitude that way before, and knew in advance of the evening's eclipse.

What had begun as an abstract geometrical study eventually had a practical payoff. The Greeks, though, scarcely could have justified their geometrical studies, had they faced an assessment procedure excluding all but those researches with evident and immediate utility.

The Greeks pursued their quest for knowledge and beauty. Civilization was the better for it, in spiritual and in material terms.

Bibliography

On Aristarchus, see Sir Thomas Heath's *Aristarchus of Samos: the Ancient Copernicus* and William Stahl's article on "The Greek Heliocentric Theory and its Abandonment", published in the *Transactions and Proceedings of the American Philological Association, volume 76* (1945), pages 321-332.

The excerpts from Archimedes' *Sand-reckoner* and from Aristarchus' *The Sizes and Distances of the Sun and Moon* are from *Aristarchus of Samos: The Ancient Copernicus* trans. Sir Thomas Heath (1913). Reprinted by permission of Oxford University Press.

The excerpt from Aristotle's *On the Heavens* is from "De Caelo" trans. J. L. Stocks, from *The Oxford Translation of Aristotle* ed. W. D. Ross. vol. 2 (1930). Reprinted by permission of Oxford University Press.

The film *Measuring Large Distances* (30 minutes) shows how to determine distance by triangulation, applies the method to the moon, sun, and stars, and compares the results to values obtained from other methods.

The film *Measuring the Universe* (11 minutes) explains stellar parallax.

The motion of the earth and the phases of the moon are shown in three short film loops *Earth: Rotation and Revolution*; *Day and Night*; and *Moon: Motion and Phases.*

CHAPTER EIGHT

Circles and More Circles

The first major attempt to solve Plato's problem for astronomers, to account for the apparently irregular planetary motions with a system of uniform circular motions, had been made by Eudoxus using homocentric spheres. The second major effort occurred not in Athens but in the new and wealthy city of Alexandria, in Egypt at the mouth of the Nile, named after Alexander the Great.

Alexander, inheriting his throne in 336 B.C., had conquered much of the Persian Empire by 331. The new empire was not yet consolidated when Alexander died suddenly in 323. Soon his generals were engaged in struggles, from which emerged three major kingdoms, in Asia, in Greece, and in Egypt. Egypt, with rich lands and a homogeneous population, enjoyed wealth and internal stability. Nor should the benefits gained by astute rulers be overlooked. It had been intended to bury Alexander in Macedonia. But, by trick or by force, the Egyptian ruler, Ptolemy I, gained possession of Alexander's body. The tomb of Alexander at Alexandria became a profitable tourist attraction. In another economic coup, Ptolemy stopped the export of grain until famine abroad brought higher prices.

Ptolemy used some of the money from these and other enterprises to found, around 290 B.C., the Museum. It became a home for as many as a hundred scholars subsidized by the government.

Not to be outdone, Ptolemy II established the Library. Its large and famous collection of books, built up from the purchase of private libraries, possibly including Aristotle's, numbered around half a million.

There may have been lectures at the Museum. Specimens of plants and animals were collected there for study. Astronomical instruments were constructed for use by philosophers at the Library. Observations and the matching of theory with fact were undertaken on a systematic and sustained basis. Galen in the biological sciences, Euclid in geometry, and Claudius Ptolemy (not one of the rulers also named Ptolemy) in astronomy all flourished in Alexandria.

Astronomers attempted to fit to the observational data systems of epicycles and eccentric circles. One of the first astronomers at Alexandria to contribute to the development of this tradition was Apollonius. He was born, probably in Athens, around 262 B.C., and moved to Alexandria at a young age. It is to Apollonius that the main credit is given for introducing the system of epicycles and eccentrics, although it is difficult to be certain what precisely were his contributions; none of his writings on astronomy have survived.

The next astronomer known to have been at Alexandria is Hipparchus, who made observations there and elsewhere around 150 B.C. References to the work of both Apollonius and Hipparchus are found in Claudius Ptolemy's book on astronomy.

All that is known of Ptolemy's life is that he was at Alexandria and made observations between the years 125 and 141 A.D. His great work on astronomy, though, has survived. It was originally called his *Mathematical Collection*, and then his *Great Collection*. Arabs later added the article *al* to the superlative *majisti*; it is by this name, the *Almagest*, that Ptolemy's work on astronomy is generally known.

Ptolemy begins his *Almagest* with a distinction between mathematics and physics, and also between these two disciplines and theology. It is the mathematical discipline that Ptolemy cultivates, particularly with respect to divine and heavenly things. He takes it as his task to describe what has already been discovered and to add his own original contributions. The motivation underlying Ptolemy's study is given by him in an epigram: "I know that I am mortal and the creature of a day; but when I search out the massed wheeling circles of the stars, my feet no longer touch the earth, but, side by side with Zeus himself, I take my fill of ambrosia, the food of the gods."

Excerpt from Ptolemy's *Almagest*

I. 1. Preface

Those who have been true philosophers, Syrus, seem to me to have very wisely separated the theoretical part of philosophy from the practical. For even if it happens the practical turns out to be theoretical prior to its being practical, nevertheless a great difference would be found in them; not only because some of the moral virtues can belong to the everyday ignorant man and it is impossible to come by the theory of whole sciences without learning, but also because in practical matters the greatest advantage is to be had from a continued and repeated operation upon the things themselves, while in theoretical knowledge it is to be had by a progress onward. We accordingly thought it up to us so to train our actions even in the application of the imagination as not to forget in whatever things we happen upon the consideration of their beautiful and well-ordered disposition, and to indulge in meditation mostly for the exposition of many beautiful theorems and especially of those specifically called mathematical.

For indeed Aristotle quite properly divides also the theoretical into three immediate genera: the physical, the mathematical, and the theological. For given that all beings have their existence from matter and form and motion, and that none of these can be seen, but only thought, in its subject separately from the others, if one should seek out in its simplicity the first cause of the first movement of the universe, he would find God invisible and unchanging. And the kind of science which seeks after Him is the theological; for such an act can only be thought as high above somewhere near the loftiest things of the universe and is absolutely apart from sensible things.

But the kind of science which traces through the material and ever moving quality, and has to do with the white, the hot, the sweet, the soft, and such things, would be called physical; and such an essence, since it is only generally what it is, is to be found in corruptible things and below the lunar sphere.

And the kind of science which shows up quality with respect to forms and local motions, seeking figure, number, and magnitude, and also place, time, and similar things, would be defined as mathematical. For such an essence falls, as it were, between the other two, not only because it can be conceived both through the senses and without the senses, but also because it is an accident in absolutely all beings both mortal and immortal, changing with those things that ever change, according to their inseparable form, and preserving unchangeable the changelessness of form in things eternal and of an ethereal nature.

And therefore meditating that the other two genera of the theoretical would be expounded in terms of conjecture rather than in terms of scientific understanding: the theological because it is in no way phenomenal and attainable, but the physical because its matter is unstable and obscure, so that for this reason philosophers could never hope to agree on them; and meditating that only the mathematical, if approached enquiringly, would give its practitioners certain and trustworthy knowledge with demonstration both arithmetic and geometric resulting from indisputable procedures, we were led to cultivate most particularly as far as lay in our power this theoretical discipline.

And especially were we led to cultivate that discipline developed in respect to divine and heavenly things as being the only one concerned with the study of things which are always what they are, and therefore able itself to be always what it is - which is indeed the proper mark of a science - because of its own clear and ordered understanding, and yet to cooperate with the other disciplines no less than they themselves.

For that special mathematical theory would most readily prepare the way to the theological, since it alone could take good aim at that unchangeable and separate act, so close to that act are the properties having to do with translations and arrangements of movements, belonging to those heavenly beings which are sensible and both moving and moved, but eternal and impassible.

Again as concerns the physical there would not be just chance correspondences. For the general property of the material essence is pretty well evident from the peculiar fashion of its local motion - for example, the corruptible and incorruptible from straight and circular movements, and the heavy and light or passive and active from movements to the center and from the center.

And indeed this same discipline [the mathematical] would more than any other prepare understanding persons with respect to nobleness of actions and character by means of the sameness, good order, due proportion, and simple directness contemplated in divine things, making its followers lovers of that divine beauty, and making habitual in them, and as it were natural, a like condition of the soul.

And so we ourselves try to increase continuously our love of the discipline of things which are always what they are, by learning what has already been discovered in such sciences by those really applying themselves to them, and also by making a small original contribution such as the period of time from them to us could well make possible. And therefore we shall try and set forth as briefly as possible as many theorems as we recognize to have come to light up to the present, and in such a way that those who have already been initiated somewhat may follow, arranging in proper order for the completeness of the treatise all matters useful to the theory of heavenly things. And in order not to make the treatise too long we shall only report what was rigorously proved by the ancients, perfecting as far as we can what was not fully proved or not proved as well as possible.

Ptolemy argues that the heavens move spherically, as is shown both by observation and by certain physical considerations, reminiscent of Aristotelian arguments. The earth is spherical, as can be inferred from a variety of observations, and the earth is in the middle of the heavens, as is also evident from observations. Likewise, the earth is very small relative to the size of the heavens.

Excerpt from Ptolemy's *Almagest*

I. 2. On the Order of the Theorems

A view, therefore, of the general relation of the whole earth to the whole of the heavens will begin this composition of ours.

And next, of things in particular, there will first be an account of the ecliptic's position and of the places of that part of the earth inhabited by us, and again of the difference, in order, between each of them according to the inclinations of their horizons. For the theory of these, once understood, facilitates the examination of the rest.

And, secondly, there will be an account of the solar and lunar movements and of their incidents. For without a prior understanding of these one could not profitably consider what concerns the stars.

The last part, in view of this plan, will be an account of the stars. Those things having to do with the sphere of what are called the fixed stars would reasonably come first, and then those having to do with what are called the five planets.

And we shall try and show each of these things using as beginnings and foundations for what we wish to find, the evident and certain appearances from the observations of the ancients and our own, and applying the consequences of these conceptions by means of geometrical demonstrations.

And so, in general, we have to state that the heavens are spherical and move spherically; that the earth, in figure, is sensibly spherical also when taken as a whole; in position, lies right in the middle of the heavens, like a geometrical center; in magnitude and distance, has the ratio of a point with respect to the sphere of the fixed stars, having itself no local motion at all. And we shall go through each of these points briefly to bring them to mind.

I. 3. That the Heavens Move Spherically

It is probable the first notions of these things came to the ancients from some such observation as this. For they kept seeing the sun and moon and other stars always moving from rising to setting in parallel circles, beginning to move upward from below as if out of the earth itself, rising little by little to the top, and then coming around again and going down in the same way until at last they would disappear as if falling into the earth. And then again they would see them, after remaining some time invisible, rising and setting as if from another beginning; and they saw that the times and also the places of rising and setting generally corresponded in an ordered and regular way.

But most of all the observed circular orbit of those stars which are always visible, and their revolution about one and the same center, led them to this spherical notion. For necessarily this point became the pole of the heavenly sphere; and the stars nearer to it were those that spun around in smaller circles, and those farther away made greater circles in their revolutions in proportion to the distance, until a sufficient distance brought one to the disappearing stars. And then they saw that those near the always-visible stars disappeared for a short time, and those farther away for a longer time proportionately. And for these reasons alone it was sufficient for them to assume this notion as a principle, and forthwith to think through also the other things consequent upon these same appearances, in accordance with the development of the science. For absolutely all the appearances contradict the other opinions.

If, for example, one should assume the movement of the stars to be in a straight line to infinity, as some have opined, how could it be explained that each star will be observed daily moving from the same starting point?

For how could the stars turn back while rushing on to infinity? Or how could they turn back without appearing to do so? Or how is it they do not disappear with their size gradually diminishing, but on the contrary seem larger when they are about to disappear, being covered little by little as if cut off by the earth's surface? But certainly to suppose that they light up from the earth and then again go out in it would appear most absurd. For if anyone should agree that such an order in their magnitudes and number, and again in the distances, places, and times is accomplished in this way at random and by chance, and that one whole part of the earth has an incandescent nature and another a nature capable of extinguishing, or rather that the same part lights the stars up for some people and puts them out for others, and that the same stars happen to appear to some people either lit up or put out and to others not yet so - even if anyone, I say, should accept all such absurdities, what could we say about the always-visible stars which neither rise nor set? Or why don't the stars which light up and go out rise and set for every part of the earth, and why aren't those which are not affected in this way always above the earth for every part of the earth? For in this hypothesis the same stars will not always light up and go out for some people, and never for others. But it is evident to everyone that the same stars rise and set for some parts, and do neither of these things for others.

In a word, whatever figure other than the spherical be assumed from the movement of the heavens, there must be unequal linear distances from the earth to parts of the heavens, wherever or however the earth be situated, so that the magnitudes and angular distances of the stars with respect to each other would appear unequal to the same people within each revolution, now larger now smaller. But this is not observed to happen. For it is not a shorter linear distance which makes them appear larger at the horizon, but the steaming up of the moisture surrounding the earth between them and our eyes, just as things put under water appear larger the farther down they are placed.

The following considerations also lead to the spherical notion: the fact that instruments for measuring time cannot agree with any hypothesis save the spherical one; that, since the movement of the heavenly bodies ought to be the least impeded and most facile, the circle among plane figures offers the easiest path of motion, and the sphere among solids; likewise that, since of different figures having equal perimeters those having the more angles are the greater, the circle is the greatest of plane figures and the sphere of solid figures, and the heavens are greater than any other body.

Moreover, certain physical considerations lead to such a conjecture. For example, the fact that of all bodies the ether has the finest and most homogeneous parts; but the surfaces of homogeneous parts must have homogeneous parts, and only the circle is such among plane figures and the sphere among solids. And since the ether is not plane but solid, it can only be spherical. Likewise the fact that nature has built all earthly and corruptible bodies wholly out of rounded figures but with heterogeneous parts, and all divine bodies in the ether out of spherical figures with homogeneous parts, since if they were plane or disc-like they would not appear circular to all those who see them from different parts of the earth at the same time. Therefore it would seem reasonable that the ether surrounding them and of a like nature be also spherical, and that because of the homogeneity of its parts it moves circularly and regularly.

I. 4. The Earth, as a Whole, is sensibly Spherical

Now, that also the earth taken as a whole is sensibly spherical, we could most likely think out in this way. For again it is possible to see that the sun and moon and the other stars do not rise and set at the same time for every observer on the earth, but always earlier for those living towards the orient [east] and later for those living towards the occident [west]. For we find that the phenomena of eclipses taking place at the same time, especially those of the moon, are not recorded at the same hours for everyone - that is, relatively to equal intervals of time from noon; but we always find later hours recorded for observers towards the orient than for those towards the occident. And since the differences in the hours is found to be proportional to the distances between the places, one would reasonably suppose the surface of the earth spherical, with the result that the general uniformity of curvature would assure every part's covering those following it proportionately. But this would not happen if the figure were any other, as can be seen from the following considerations.

For, if it were concave, the rising stars would appear first to people towards the occident; and if it were flat, the stars would rise and set for all people together and at the same time; and if it were a pyramid, a cube, or any other polygonal figure, they would again appear at the same time for all observers on the same straight line. But none of these things appears to happen. It is further clear that it could not be cylindrical with the curved surface turned to the risings and settings and the plane bases to the poles of the universe, which some think more plausible. For then never would any of the stars be always visible to any of the inhabitants of the curved surface, but either all the stars would both rise and set for observers or the same stars for an equal distance from either of the poles would always be invisible to all observers. Yet the more we advance toward the north pole, the more the southern stars are hidden and the northern stars appear. So it is clear that here the curvature of the earth covering parts uniformly in oblique directions proves its spherical form on every side. Again, whenever we sail towards mountains or any high places from whatever angle and in whatever direction, we see their bulk little by little increasing as if they were arising from the sea, whereas before they seemed submerged because of the curvature of the water's surface.

I. 5. The Earth is in the Middle of the Heavens

Now with this done, if one should next take up the question of the earth's position, the observed appearances with respect to it could only be understood if we put it in the middle of the heavens as the center of the sphere. If this were not so, then the earth would either have to be off the axis but equidistant from the poles, or on the axis but farther advanced towards one of the poles, or neither on the axis nor equidistant from the poles.

The following considerations are opposed to the first of these three positions - namely, that if the earth were conceived as placed off the axis either above or below in respect to certain parts of the earth, those parts, in the right sphere, would never have any equinox since the section above the earth and the section below the earth would always be cut unequally by the horizon. Again, if the sphere were inclined with respect to these parts, either they would have no equinox or else the equinox would not take place midway between the summer and winter solstices. The distances would be unequal because the equator which is the greatest of those parallel circles described about the poles would not be cut in half by the horizon; but one of the circles parallel to it, either to the north or to the south, would be so cut in half. It is absolutely agreed by all, however, that these distances are everywhere equal because the increase from the equinox to the longest day at the summer tropic are equal to the decreases to the least days at the winter tropic. And if the deviation for certain parts of the earth were supposed either towards the orient or the occident, it would result that for these parts neither the sizes and angular distances of the stars would appear equal and the same at the eastern and western horizons, nor would the time from rising to the meridian be equal to the time from the meridian to setting. But these things evidently are altogether contrary to the appearances.

As to the second position where the earth would be on the axis but farther advanced towards one of the poles, one could again object that, if this were so, the plane of the horizon in each latitude would always cut into uneven parts the sections of the heavens below the earth and above, different with respect to each other and to themselves for each different deviation. And the horizon could cut into two even parts only in the right sphere. But in the case of the inclined sphere with the nearer pole ever visible, the horizon would always make the part above the earth less and the part below the earth greater with the result that also the great circle through the center of the signs of the zodiac [ecliptic] would be cut unequally by the plane of the horizon.

But this has never been seen, for six of the twelve parts are always and everywhere visible above the earth, and the other six invisible; and again when all these last six are all at once visible, the others are at the same time invisible. And so - from the fact that the same semicircles are cut off entirely, now above the earth, now below - it is evident that the sections of the zodiac are cut in half by the horizon.

And, in general, if the earth did not have its position under the equator but lay either to the north or south nearer one of the poles, the result would be that, during the equinoxes, the shadows of the gnomons at sunrise would never perceptibly be on a straight line with those at sunset in planes parallel to the horizon. But the contrary is everywhere seen to occur. And it is immediately clear that it is not possible to advance the third position since each of the obstacles to the first two would be present here also.

In brief, all the observed order of the increases and decreases of day and night would be thrown into utter confusion if the earth were not in the middle. And there would be added the fact that the eclipses of the moon could not take place for all parts of the heavens by a diametrical opposition to the sun, for the earth would often not be interposed between them in their diametrical oppositions, but at distances less than a semicircle.

I. 6. Earth's Ratio is as a Point to the Heavens

Now, that the earth has sensibly the ratio of a point to its distance from the sphere of the so-called fixed stars gets great support from the fact that in all parts of the earth the sizes and angular distances of the stars at the same times appear everywhere equal and alike, for the observations of the same stars in the different latitudes are not found to differ in the least.

Moreover, this must be added: that sundials placed in any part of the earth and the centers of armillary spheres can play the role of the earth's true center for the sightings and the rotations of the shadows, as much in conformity with the hypotheses of the appearances as if they were at the true midpoint of the earth.

And the earth is clearly a point also from this fact: that everywhere the planes drawn through the eye, which we call horizons, always exactly cut in half the whole sphere of the heavens. And this would not happen if the magnitude of the earth with respect to its distance from the heavens were perceptible; but only the plane drawn through the point at the earth's center would exactly cut the sphere in half, and those drawn through any other part of the earth's surface would make the sections below the earth greater than those above.

Ptolemy accepts Aristotle's ideas on motion, including a natural place for each of the various elements. The earth cannot change position, both because such movement would then be observed in apparent changes in the positions of the stars and because the earth's natural place is at the center of the universe. Nor can the earth rotate on its axis, for reasons which Ptolemy sets forth and which were debated over for centuries. One of the arguments was that if the earth were rotating, would not the clouds or anything thrown into the air then be seen to be left behind as the earth rotated under it?

Excerpt from Ptolemy's *Almagest*

I. 7. That the Earth Does Not in any Way Move Locally

By the same arguments as the preceding it can be shown that the earth can neither move in any one of the aforesaid oblique directions, nor ever change at all from its place at the center. For the same things would result as if it had another position than at the center. And so it also seems to me superfluous to look for the causes of the motion to the center when it is once for all clear from the very appearances that the earth is in the middle of the world and all weights move towards it. And the easiest and only way to understand this is to see that, once the earth has been proved spherical considered as a whole and in the middle of the universe as we have said, then the tendencies and movements of heavy bodies (I mean their proper movements) are everywhere and always at right angles to the tangent plane drawn through the falling body's point of contact with the earth's surface. For because of this it is clear that, if they were not stopped by the earth's surface, they too would go all the way to the center itself, since the straight line drawn to the center of a sphere is always perpendicular to the plane tangent to the sphere's surface at the intersection of that line.

All those who think it paradoxical that so great a weight as the earth should not waver or move anywhere seem to me to go astray by making their judgment with an eye to their own affects and not to the property of the whole. For it would not still appear so extraordinary to them, I believe, if they stopped to think that the earth's magnitude compared to the whole body surrounding it is in the ratio of a point to it. For thus it seems possible for that which is relatively least to be supported and pressed against from all sides equally and at the same angle by that which is absolutely greatest and homogeneous. For there is no "above" and "below" in the universe with respect to the earth, just as none could be conceived of in a sphere. And of the compound bodies in the universe, to the extent of their proper and natural motion, the light and subtle ones are scattered in flames to the outside and to the circumference, and they seem to rush in the upward direction relative to each one because we too call "up" from above our heads to the enveloping surface of the universe; but the heavy and coarse bodies move to the middle and center and they seem to fall downwards because again we all call "down" the direction from our feet to the earth's center. And they properly subside about the middle under the everywhere-equal and like resistance and impact against each other. Therefore the solid body of the earth is reasonably considered as being the largest relative to those moving against it and as remaining unmoved in any direction by the force of the very small weights, and as it were absorbing their fall.

And if it had some one common movement, the same as that of the other weights, it would clearly leave them all behind because of its much greater magnitude. And the animals and other weights would be left hanging in the air, and the earth would very quickly fall out of the heavens. Merely to conceive such things makes them appear ridiculous.

Now some people, although they have nothing to oppose to these arguments, agree on something, as they think, more plausible. And it seems to them there is nothing against their supposing, for instance, the heavens immobile and the earth as turning on the same axis from west to east very nearly one revolution a day; or that they both should move to some extent, but only on the same axis as we said, and conformably to the overtaking of the one by the other.

But it has escaped their notice that, indeed, as far as the appearances of the stars are concerned, nothing would perhaps keep things from being in accordance with this simpler conjecture, but that in the light of what happens around us in the air such a notion would seem altogether absurd. For in order for us to grant them what is unnatural in itself, that the lightest and subtlest bodies either do not move at all or no differently from those of contrary nature, while those less light and less subtle bodies in the air are clearly more rapid than all the more terrestrial ones; and to grant that the heaviest and most compact bodies have their proper swift and regular motion, while again these terrestrial bodies are certainly at times not easily moved by anything else - for us to grant these things, they would have to admit that the earth's turning is the swiftest of absolutely all the movements about it because of its making so great a revolution in a short time, so that all those things that were not at rest on the earth would seem to have a movement contrary to it, and never would a cloud be seen to move toward the east nor anything else that flew or was thrown into the air. For the earth would always outstrip them in its eastward motion, so that all other bodies would seem to be left behind and to move towards the west.

For if they should say that the air is also carried around with the earth in the same direction and at the same speed, nonetheless the bodies contained in it would always seem to be outstripped by the movement of both. Or if they should be carried around as if one with the air, neither the one nor the other would appear as outstripping, or being outstripped, by, the other. But these bodies would always remain in the same relative position and there would be no movement or change either in the case of flying bodies or projectiles. And yet we shall clearly see all such things taking place as if their slowness or swiftness did not follow at all from the earth's movement.

I. 8. That There Are Two Different Prime Movements in the Heavens

It will be sufficient for these hypotheses, which have to be assumed for the detailed expositions following them, to have been outlined here in such a summary way since they will finally be established and confirmed by the agreement of the consequent proofs with appearances. In addition to those already mentioned, this general assumption would also be rightly made that there are two different prime movements in the heavens. One is that by which everything moves from east to west, always in the same way and at the same speed with revolutions in circles parallel to each other and clearly described about the poles of the regularly revolving sphere. Of these circles the greatest is called the equator, because it alone is always cut exactly in half by the horizon which is a great circle of the sphere, and because everywhere the sun's revolution about it is sensibly equinoctial. The other movement is that according to which the spheres of the stars [planets] make certain local motions in the direction opposite [retrograde] to that of the movement just described and around other poles than those of that first revolution. And we assume that it is so because, while, from each day's observation, all the heavenly bodies are seen to move generally in paths sensibly similar and parallel to the equator and to rise, culminate, and set (for such is the property of the first movement), yet from subsequent and more continuous observation, even if all the other stars appear to preserve their angular distances with respect to each other and their properties as regards their places within the first movement, still the sun and moon and planets make certain complex movements unequal to each other, but all contrary to the general movement, towards the east opposite to the movement of the fixed stars which preserve their respective angular distances and are moved as if by one sphere.

If, then, this movement of the planets also took place in circles parallel to the equator - that is, around the same poles as those of the first revolution - it would be sufficient to assume for them all one and the same revolving movement in conformity with the first. For it would then be plausible to suppose that their movement was the result of a lag and not of a contrary movement. But they always seem, at the same time they move towards the east, to deviate towards the north and south poles without any uniform magnitudes being observed in this deviation, so that this seems to befall them through impulsions. But although this deviation is irregular on the hypothesis of one prime movement, it is regular when effected by a circle oblique to the equator. And so such a circle is conceived one and the same for, and proper to, the planets, quite exactly expressed and as it were described by the motion of the sun, but traveled also by the moon and planets which ever turn about it with every deviation from it on the part of any planet either way, a deviation within a prescribed distance and governed by rule. And since this is seen to be a great circle also because of the sun's equal oscillation to the north and south of the equator, and since the eastward movements of all the planets (as we said) take place on one and the same circle, it was necessary to suppose a second movement different from the general one, a movement about the poles of this oblique circle or ecliptic in the direction opposite to that of the first movement.

Then if we think of a great circle described through the poles of both the circles just mentioned, which necessarily cuts each of them - that is, the equator and the circle inclined to it - exactly in half and at right angles, there will be four points on the oblique circle or ecliptic: the two made by the equator diametrically opposite each other and called the equinoxes of which the one guarding the northern appproach is called spring, and the opposite one autumn. And the two made by the circle drawn through both sets of poles, also clearly diametrically opposite each other, are called the tropics, of which the one to the south of the equator is called winter, and the one to the north summer.

The one first movement which contains all the others will be thought of then as described and as if defined by the great circle, through both sets of poles, which is carried around and carries with it all the rest from east to west about the poles of the equator. And these poles are as if they were on what is called the meridian, which differs from the circle through both sets of poles in this alone: that it is not always drawn through the poles of the ecliptic, but is conceived as continuously at right angles to the horizon and therefore called the meridian, since such a position cutting in half as it does each of the two hemispheres, that below the earth and that above, provides midday and midnight. But the second movement, consisting of many parts and contained by the first, and embracing itself all the planetary spheres, is carried by the first as we said, and revolves about the poles of the ecliptic in the opposite direction. And these poles of the ecliptic bring on the circle effecting the first revolution - that is, on the circle drawn through all four poles together - are carried around with it as one would expect; and, moving therefore with a motion opposite to the second prime movement, in this way keep the position of the great circle which is the ecliptic ever the same with respect to the equator.

I. 9. On the Particular Motions

A summary and general preliminary explanation would contain some such exposition as the foregoing of the things to be presupposed. But now we are going to begin the detailed proofs. And we think the first of these is that by means of which is calculated the length of the arc between the poles of the equator and the ecliptic, lying on the great circle drawn through these poles. To this end we must first see expounded the method of computing the size of chords inscribed in a circle, and we are now going to demonstrate this geometrically for each case, once for all.

Ptolemy concludes Book I with a summary of the motions of the planets and a long development of trigonometric techniques necessary for a detailed use of astronomical measurements. Book II is devoted to the application of the trigonometric techniques to an oblique sphere, such as the earth.

The Eccentric Hypothesis

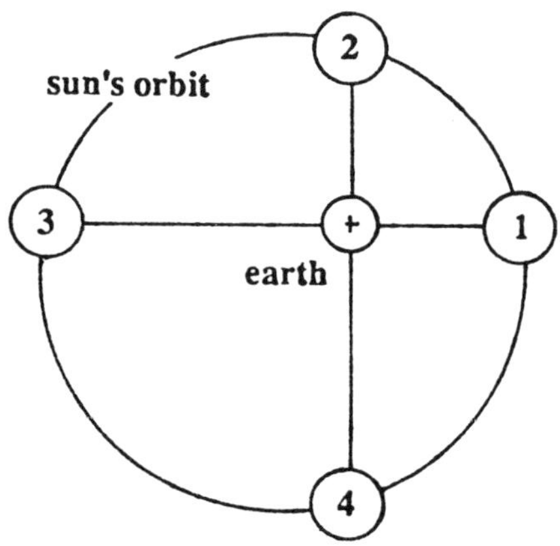

The sun, moving in its orbit with constant speed (traversing equal distances in equal times), viewed from the earth moves through the 90 degree angle from 1 to 2 faster (in less time) than it moves through the 90 degree angle from 2 to 3. The journey from 3 to 4, cutting off another 90 degree angle, consumes even more time. A constant angular speed relative to the center of the orbit thus appears irregular when viewed from any point other than the center.

In book III, Ptolemy takes up the problem of the sun's motion. He describes previous observations of the length of the year and summarizes in a table the results of the sun's regular movement. The next task, he writes, is to explain the apparent irregularity of the sun's motion as a combination of regular circular motions.

Regular circular motion is defined as that motion which cuts off equal angles in equal times at the center of the circle. Difficulties in matching theory to observation forced Ptolemy to violate his own definition of uniform circular motion. The violation became one of the major causes of dissatisfaction with Ptolemy's system, leading to Copernicus' revision and revolution fourteen centuries later.

The Epicycle Hypothesis

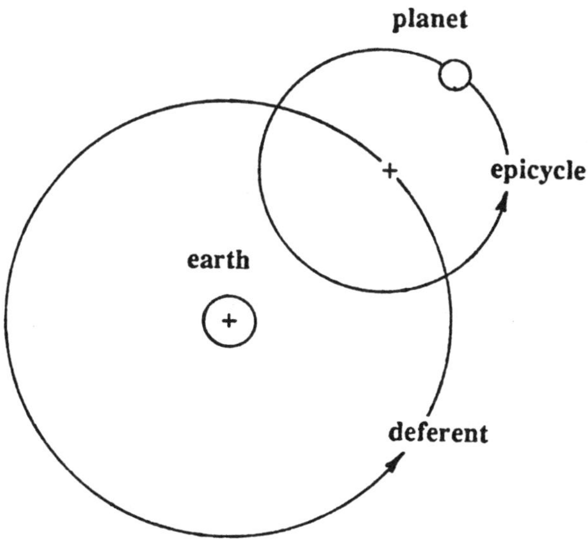

The deferent circle rotates around the earth. The center of the epicycle circle is attached to and carried around by the deferent. The epicycle rotates about its center. The planet is attached to and carried around by the epicycle.

In his discussion of the sun's motion, Ptolemy describes both the eccentric hypothesis and the epicycle theory. In the former, the circle carrying the sun is not centered on the earth, and regular motion as viewed from the center of the sun's orbit appears irregular when viewed from the earth; the sun viewed from the earth appears to move with varying speeds.

In the epicycle theory, a small circle (the epicycle) moves on a large circle (the deferent) and the combination of their regular motions is irregular.

In the case of the sun, either hypothesis can produce the observed motion. The hypotheses are interchangeable in a mathematical sense, though not in an Aristotelian physical sense. To produce the more complex movements of the planets, Ptolemy found it necessary to employ both hypotheses simultaneously.

Excerpt from Ptolemy's *Almagest*

III. 3. On the Hypotheses concerning Regular and Circular Movement

Since the next thing is to explain the apparent irregularity of the sun, it is first necessary to assume in general that the motions of the planets in the direction contrary to the movement of the heavens are all regular and circular by nature, like the movement of the universe in the other direction. That is, the straight lines, conceived as revolving the stars on their circles, cut off in equal times on absolutely all circumferences equal angles at the centers of each, and their apparent irregularities result from the positions and arrangements of the circles on their spheres through which they produce these movements, but no departure from their unchangeableness has really occurred in their nature in regard to the supposed disorder of their appearances.

But the cause of this irregular appearance can be accounted for by as many as two primary simple hypotheses. For if their movement is considered with respect to a circle in the plane of the ecliptic concentric with the cosmos so that our eye is the center, then it is necessary to suppose that they make their regular movements either along circles not concentric with the cosmos, or along concentric circles; not with these simply, but with other circles borne upon them called epicycles. For according to either hypothesis it will appear possible for the planets seemingly to pass, in equal periods of time, through unequal arcs of the ecliptic circle which is concentric with the cosmos.

For if, in the case of the hypothesis of eccentricity, we conceive the eccentric circle ABCD *on which the star moves regularly, with* E *as center and with diameter* AED, *and the point* F *on it as your eye so that the point* A *becomes the apogee [farthest from the earth] and the point* D *the perigee [closest to the earth]; and if, cutting off equal arcs* AB *and* DC, *we join* BE, BF, CE, *and* CF, *then it will be evident that the star moving through each of the arcs* AB *and* CD *in an equal period of time will seem to have passed through unequal arcs on the circle described around* F *as center. For since angle* BEA *equals angle* CED, *therefore angle* BFA *is less than either of them, and angle* CFD *greater.*

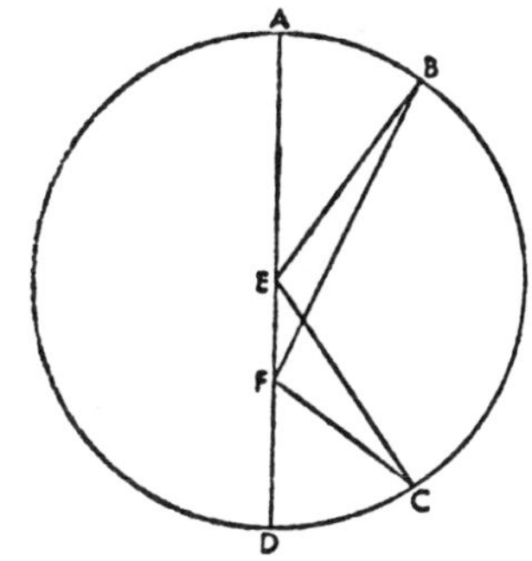

And if in the hypothesis of the epicycle we conceive the circle ABCD *concentric with the ecliptic with center* E *and diameter* AEC, *and the epicycle* FGHK *borne on it on which the star moves, with its center at* A, *then it will be immediately evident also that as the epicycle passes regularly along the circle* ABCD, *from* A *to* B *for example, and the star along the epicycle, the star will appear indifferently to be at* A *the center of the epicycle when it is at* F *or* H: *but when it is at other points, it will not. But having come to* G, *for instance, it will seem to have produced a movement greater than the regular movement by the arc* AG; *and having come to* K, *likewise less by the arc* AK.

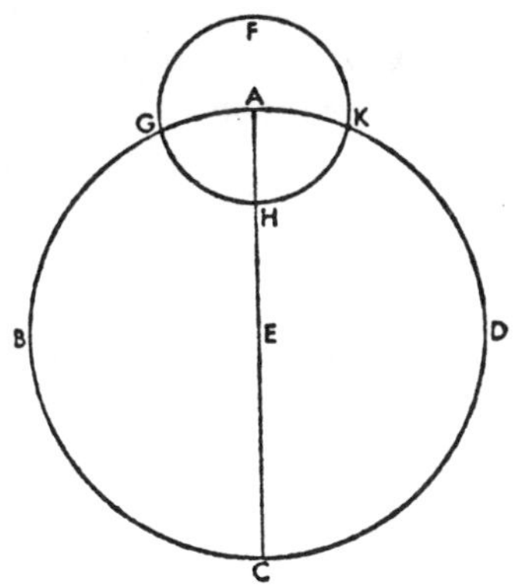

Then with the hypothesis of eccentricity it is always the case that the least movement belongs to the apogee and the greatest movement to the perigee, since angle AFB is *always less than angle* DFC. *But both cases can come about with the hypothesis of the epicycle. For when the epicycle moves contrary to the heavens [from west to east], for example from* A *to*

B, *if the star so moves on the epicycle that it goes from the apogee again contrary to the heavens (that is, from* F *in the direction of* G*), there will result at the apogee the greatest advance, because the epicycle and the star are moving the same way. But if the movement of the star on the epicycle is in the direction of that of the heavens [from east to west], that is, from* F *towards* K, *conversely the least advance will be effected at the apogee because the star is then moving contrary to the movement of the epicycle.*

With these things established, it must next be understood that, in the case of those planets which effect two anomalies, it is possible to combine both of these hypotheses, as we shall show in the chapters concerning them. But, in the case of those planets subject to only one anomaly, one of the hypotheses will suffice. And it must be understood that all the appearances can be cared for interchangeably according to either hypothesis, when the same ratios are involved in each. In other words, the hypotheses are interchangeable when, in the case of the hypothesis of the epicycle, the ratio of the epicycle's radius to the radius of the circle carrying it is the same as, in the case of the hypothesis of eccentricity, the ratio of the line between the centers (that is, between the eye and the center of the eccentric circle), to the eccentric circle's radius; with the added conditions that the star move on the epicycle from the apogee in the direction of the movement of the heavens with the same angular velocity as the epicycle moves on the circle concentric with the eye in the direction opposite to that of the heavens, and that the star move regularly on the eccentric circle with the same angular velocity also and in the direction opposite to the movement of the heavens.

And we shall briefly show in a systematic way, first by reasoning and secondly by the numbers discovered in the appearances of the sun's anomaly, that with the above assumptions the same appearances agree with either hypothesis.

Book IV of the *Almagest* is concerned with the motions of the moon. Ptolemy presents a table of the motions of the moon, shows that the eccentric circle and epicycle hypotheses produce the same appearances, and develops a model to match in detail the appearances. There were a variety of complicating features in the motion of the moon. Furthermore, the lunar motion had to be linked with the motion of the sun if eclipses were to be accounted for.

Both Hipparchus and Ptolemy had made, or at least they reported, observations of greater accuracy than had ever been made before. They used an instrument called the astrolabe, which Ptolemy describes in the *Almagest* at the beginning of Book V. This exposition is followed by further development of the lunar theory, which also comprises Book VI.

Examinations of Ptolemy's reported observations reveal that many of them were fabricated; they were calculated from theory rather than actually observed. It is as if a student were assigned a laboratory experiment, knew what the answer should be, and reported that value without doing the experiment. Under modern scientific ethics, Ptolemy is guilty of a serious crime. Indeed, his book has been called the greatest scientific fraud of all time.

In partial defense of Ptolemy, it may be noted that he must have had a large number of reasonably good observations, because his theory does agree well with the facts. Possibly Ptolemy was presenting more in the spirit of example than of proof "corrected" numerical values, calculated from a theory which in turn was based on a large number of observations not mentioned by Ptolemy.

Ptolemy moves on to the stars in Books VII and VIII, and then to the planets in the last five books. He accepts the traditional order of the planets, their relative distances from the earth, and restates the basic problem of devising a system of regular circular motions. The task will be difficult, Ptolemy writes, and he may be forced to use something contrary to the general argument or to presuppose something without immediate foundation or to devise different sorts of models for the different planets.

Excerpt from Ptolemy's *Almagest*

IX. The Order of the Spheres of the Sun and Moon and Five Planets

Now, certainly whatever one could say in general about the fixed stars, to the extent that the appearances up until now fall under our apprehension, would be pretty much like this. But since this Composition still lacks a treatment of the five planets, we shall give an exposition of them, going as far as possible with what they have in common to avoid repetition, and then adding on the plan of each one in particular.

First, then, concerning the order of their spheres, all of which have their positions about the poles of the ecliptic, we see the foremost mathematicians agree that all these spheres are nearer the earth than the sphere of the fixed stars, and farther from the earth than that of the moon; that the three - of which Saturn's is the largest, Jupiter's next earthward, and Mars' below that - are all farther from the earth than the others and that of the sun. On the other hand, the spheres of Venus and Mercury are placed by the earlier mathematicians below the sun's, but by some of the later ones above the sun's because of their never having seen the sun eclipsed by them. But this judgment seems to us unsure since these planets could be below the sun and never yet have been in any of the planes through the sun and our eye but in another, and therefore not have appeared in a line with it; just as in the case of the moon's conjunctive pasages there are for the most part no eclipses.

Since there is no other way of getting at this because of the absence of any sensible parallax in these stars, from which appearance alone linear distances are gotten, the order of the earlier mathematicians seems the more trustworthy, using the sun as a natural dividing line between those planets which can be any angular distance from the sun and those which cannot but which always move near it. Besides, it does not place them far enough at their perigees to produce a sensible parallax.

IX. 2. On the Aim of the Planetary Hypothesis

So much, then, for the orders of the spheres. Now, since our problem is to demonstrate, in the case of the five planets as in the case of the sun and moon, all their apparent irregularities as produced by means of regular and circular motions (for these are proper to the nature of divine things which are strangers to disparities and disorder) the successful accomplishment of this aim as truly belonging to mathematical theory in philosophy is to be considered a great thing, very difficult and as yet unattained in a reasonable way by anyone. For, since, in the case of the researches about the periodic movements of each planet, whatever slight error the eye makes in systematic observations produces a sensible difference more quickly when the examination has been over a shorter interval than when over a greater one, the time for which we have observations of the planets recorded, being short for grasping such a considerable lay-out, furnishes an unsure prediction over long periods. In the case of research about the anomalies, the fact that there are two anomalies appearing for each of the planets, and that they are unequal in magnitude and in the times of their returns, works a good deal of confusion. For one of the anomalies is seen to have relation to the sun, and the other to the parts of the zodiac, but both are mixed together so it is very hard to determine what belongs to each; and most of the old observations were thrown together carelessly and grossly. The more continuous of them contain stations and apparitions, and the apprehension of these properties is not certain. The stations cannot indicate the exact time, since the planet's local motion remains imperceptible for many days before and after its station; and the apparitions not only make the places immediately disappear along with the stars as they are seen for the first or last time, but also can be utterly misleading as to the times because of the differences in the atmosphere and in the eye of the observer. In general, the observations made with reference to some fixed star at a rather great angular distance, unless because of these things one attends to them wisely and clear-sightedly, furnish a magnitude from their measurements hard to calculate and subject to guesswork. And this is so, not only because the lines between the observed stars make different angles with the ecliptic and by no means right angles - whence in the variety of the zodiac's inclinations a great uncertainty is apt to follow in the determination of the longitudinal and latitudinal positions - but also because the same angular distances appear to the eye greater near the horizon and smaller near the culminations, and so they can be measured as sometimes greater and sometimes smaller than the real angular distance.

And so I consider Hipparchus to have been most zealous after the truth, both because of all these things and especially because of his having left us more examples of accurate observations than he ever got from his predecessors. He sought out the hypotheses of the sun and moon, and demonstrated as far as possible and by every available means that they were accomplished through uniform circular movements, but he did not attempt to give the principle of the hypotheses of the five planets, as far as we can tell from those memoirs of his which have come down to us, but only arranged the observations in a more useful way and showed the appearances to be inconsistent with the hypotheses of the mathematicians of that time.

For not only did he think it necessary as it seemed to declare that, because of the double anomaly of each planet, the regressions of each are unequal and of such and such a magnitude, while the other mathematicians gave their geometrical demonstrations on one and the same anomaly and regression, but he also thought that these movements could not be effected either by eccentric circles, or by circles concentric with the ecliptic but bearing epicycles, or even by both together, although the zodiacal anomaly was on one magnitude and the anomaly with respect to the sun of another. For these are the means used by nearly all those who have wished to demonstrate uniform circular movement by the so-called perpetual table, but in a false and inconsequential way, some getting nowhere at all, some following the problem to a limited extent.

But Hipparchus reasoned that no one who has progressed through the wole of mathematics to such a point of accuracy and zeal for truth would be content to stop at this like the rest; but that anyone who was to persuade himself and those in touch with him would have to demonstrate the magnitude and periods of each of the anomalies by clear and consistent appearances; and, putting both together, he would have to find out the position and order of the circles by which these anomalies are produced and the mode of their movement and finally show about all the appearances to be consistent with the peculiar property of this hypothesis of the circles.

I think this is difficult, and it seemed so to him. We have said all this not through ostentation, but in order that, if we are forced by the problem itself either (1) to use something contrary to the general argument, as when, for example, for ease we make our demonstrations of the circles described by the movement in the planetary spheres as if they were simple and in the same plane with the ecliptic; of if we are forced (2) to presuppose something without immediate foundation in the appearances, an apprehension gotten from continuous trial and adjustment; or (3) to suppose not everywhere the same mode of movement or inclination of the circles - in order that, I say, we may then reasonably agree that (1) using something of the sort that no appreciable difference is to result from it will not falsify the subject in hand; (2) that things supposed without proof, once they are conceived in such a way as to be consistent with appearances, cannot be found without some plan and knowledge even if the way of getting hold of them is hard to explain (after all, generally speaking, the cause of first principles is either nothing or hard to interpret in its nature): and (3) that, since the appearances relative to the stars are also found to be dissimilar, one should not reasonably think it strange or absurd to vary the mode of the hypotheses of the circles, especially when, along with saving the regular circular movement absolutely everywhere, each of the appearances is demonstrated in its more lawful and general character.

Ptolemy next presents tables of the planets' motions and his preliminary hypotheses for saving the phenomena. He describes the apparent differences from regular motion, the anomalies. Then Ptolemy reports that things are not proceeding simply.

One complication involves the planes of the eccentric circles. To produce the observed motions of the planets, Ptolemy finds that the planes of the eccentric circles cannot be immobile.

A more serious complication occurs when Ptolemy finds it necessary to modify his definition of regular motion in order to save the phenomena. He has the epicycles move with motions regular with respect to a point other than at the center of the eccentric circle which carries them. The imaginary point about which regular circular motion takes place is the equant point. It is this addition to the planetary theory, motion about a point not at the center of the earth, that Copernicus will see, fourteen centuries later, as an unacceptable violation of the rule of regular circular motion.

Another, perhaps less important, difficulty with Ptolemy's system related to the equant point was that the position of the equant point was different for Mercury than for the other planets. This was a violation of symmetry. The violation would disturb the sensibilities of later astronomers, regardless of the good agreement of Ptolemy's system with obsevations. Indeed, it was Ptolemy's insistence upon accurately accounting for all the known motions that forced him to compromise on the aesthetics of his system.

Among the observations creating difficulties were variations in the latitudes of the planets, for which Ptolemy devised little circles to vary the inclinations of the eccentric circles of the planets. He realized that his readers would react unfavorably to these and other devices. Ptolemy pleaded that when it was not possible to fit the simpler hypotheses to the movements in the heavens, it was proper to try any hypotheses. Simplicity is another aesthetic aspect of scientific theories whose importance should not be underestimated.

Excerpt from Ptolemy's *Almagest*

IX. 5. Preliminaries of the Hypotheses of the Five Planets

Now, the relation of the anomalies to the longitudinal passage of the five planets follows the exposition of these mean movements, and we have attempted a general outline of it in the following way.

For, as we said, the very simple movements together sufficient for the problem in hand are two: one effected by circles eccentric to the ecliptic, and the other by circles concentric with the ecliptic but bearing epicycles.

And likewise also the apparent anomalies for each star considered singly are two: one observed with respect to the parts of the zodiac and the other with respect to the configurations of the sun.

In the case of this latter anomaly, we find, from different configurations observed in contiguity and in the same parts of the zodiac, that, for the five planets, the time from the greatest movement to the mean movement is always longer than from the mean to the least. And such a property cannot follow from the hypothesis of eccentricity, but its contrary follows, because the greatest passage is always effected at the perigee, and in both hypotheses the arc from the perigee to the point of mean passage is less than that from this point to the apogee.

But it can occur in the hypothesis of epicycles when the greatest passage is not effected at the perigee as in the case of the moon, but at the apogee - that is, when the star, starting from the apogee, moves not westward as the moon, but eastward in the opposite direction. And so we suppose this anomaly to be produced by epicycles.

But in the case of the anomaly observed with respect to the parts of the zodiac, we find, by means of the arcs of the zodiac taken at the same phases or configurations, that, on the contrary, the time from the least movement to the mean is always greater than the time from the mean movement to the greatest. But this property can follow from either hypothesis, and at the beginning of the composition of the sun we showed how they were alike in this. But since it is more proper to the eccentric hypothesis, we suppose this anomaly is effected according to it, and also because the other anomaly is peculiar to the epicycle hypothesis.

But immediately on applying the particular positions observed to the courses constructed from the combination of both hypotheses and continually examining them together, we find things cannot proceed so simply:
1. The planes in which we describe the eccentric circles are not immobile, so that the straight line through both their centers and the ecliptic's center, along which the apogees and perigees are sighted, always remains at the same angular distances from the tropic and equinoctial points.
2. The epicycles do not have their centers borne on the eccentric circles whose centers are those with respect to which the epicycles' centers revolve

in a regular eastward motion and cut off equal angles in equal times.
But:
1. The eccentrics' apogees make a slight regular shift eastward from the tropic points around the ecliptic's center and nearly as much for each planet as the sphere of the fixed stars is found to make - that is, one degree in a hundred years, as far as one can detect from present data. And
2. The epicycles' centers are borne on circles equal to the eccentrics effecting the anomaly, but described about other centers.

And these other centers, in the case of all except Mercury, bisect the straight lines between the centers of the eccentrics effecting the anomaly and the center of the ecliptic. But in the case of Mercury alone, this other center is the same distance from the center revolving it [the equant point] as this center revolving it is in turn from the center effecting the anomaly on the side of the apogee, and as this last center effecting the anomaly is in turn from the center placed at the eye.

For, in the case of that star [Mercury] alone as also with the moon, we find the eccentric circle revolved by the aforesaid center, contrariwise to the epicycle, back westward one revolution in a year's time, since it appears to be twice perigee in one revolution, just as the moon is also twice so in one month's time.

XIII. 2. On the Manner of the Movement of the Obliquities and Slantings

The general conclusion of the hypotheses is, then, that the eccentric circles of the five planets are inclined to the ecliptic's plane about the ecliptic's center: without movement in the case of the three, Saturn, Jupiter, and Mars, so that diametrically opposite passages of the epicycles are borne in contrary latitudinal directions; but in the case of Venus and Mercury, changing simultaneously with the epicycles towards the same latitude, always towards the north for Venus and always towards the south for Mercury; and that those diameters of the epicycles through their apparent apogees, starting somewhere in the eccentric's plane, are carried aside by little circles lying beside the perigeal limits, say, proportionate in size to the latitudinal deviation, perpendicular to the eccentrics' planes, and having their centers on them. These little circles are revolved regularly and in accord with the longitudinal passages starting from an intersection of the planes with the epicycles (say towards the north) and at the same time carrying the epicycles' planes in the first quadrant's turn to the northernmost limit, of course, then back again to the eccentric's plane; and in the third quadrant's turn to the southernmost limit; and, finally, at the return to the starting point. The beginning and return of this circuit are established - for Saturn, Jupiter, and Mars - from the intersection at the ascending node; in the case of Venus, from the eccentric's perigee; and in the case of Mercury, from the eccentric's apogee.

In the case of the three stars, the diameters perpendicular to these first diameters remain, as we said, always parallel to the ecliptic's plane or have an inappreciable slant to it. But in the case of Mercury and Venus, starting again from some point in the ecliptic's plane, they are carried aside by little circles lying beside, say, their eastern limits, again proportionate in size to the latitudinal deviation, perpendicular to the ecliptic's plane, and having their centers on diameters parallel to the ecliptic's plane. These little circles are revolved at the same speed as the others starting from an intersection of the planes and the epicycles again northwards (let us say), carrying the eastern extremities of these diameters in the same order as before. And again for these stars, the beginning and return of the similar circuit are established, for Venus, from the node in the additive semicircle, and, for Mercury, from the node in the subtractive semicircle.

Now, concerning these little circles by which the oscillations of the epicycles are effected, it is necessary to assume that they are bisected by the planes about which we say the swayings of the obliquities take place; for only in this way can equal latitudinal passages be established on either side of them. Yet they do not have their revolutions with respect to regular movement effected about the proper center, but about another which has the same eccentricity for the little circle as the star's longitudinal eccentricity for the ecliptic. For the returns on the ecliptic and the little circle are supposed isochronous and the quarterly passages on each apparently agree. But if the little circle's revolutions take place about its proper center, that which is required can in no way come about, since the passages for the little circle travel each quadrant in equal time, but the epicycle's passages with respect to the ecliptic do not because of the eccentricity assumed in each case. But if they take place about a center similar in position to that of the eccentric, and of the quadrants, the returns of the obliquities will traverse corresponding parts of the ecliptic and of the little circle in equal times.

Let no one, seeing the difficulty of our devices, find troublesome such hypotheses. For it is not proper to apply human things to divine things nor to get beliefs concerning such great things from such dissimilar examples. For what is more unlike than those which are always alike with respect to those which never are, and than those which are impeded by anything with those which are not even impeded by themselves? But it is proper to try and fit as far as possible the simpler hypotheses to the movements in the heavens; and if this does not succeed, then any hypotheses possible. Once all the appearances are saved by the consequences of the hypotheses, why should it seem strange that such complications can come about in the movements of heavenly things? For there is no impeding nature in them, but one proper to the yielding and giving way to movements according to the nature of each planet, even if they are contrary, so that they can all penetrate and shine through absolutely all the fluid media; and this free action takes place not only about the particular circles, but also about the spheres themselves and the axes of revolution. We see the complication and sequence in their different movements difficult and hard to come by for the freedom of the movements in the likely stories constructed by us, but in the heavenly thing never anywhere impeded by this mixture. Or rather it is not proper to judge the simplicity of heavenly things by those which seem so with us, when here not even to all of us does the same thing seem likewise simple. For in this way not one of the heavenly occurrences would seem simple to those studying them, not even the unchangeableness of the first motion, since always to be the same is not difficult here with us but impossible. We should instead judge their simplicity from the unchangeableness of the natures in the heavens and their movements. For thus they would all appear simple, more than those things which seem so here with us, since no difficulty or trouble could be conjectured concerning their periods.

There were difficulties with Ptolemy's system, especially in its minute details. In a broad qualitiative sense, however, the epicycle system seemed a promising approach to the problem of the planets. The epicycle system accounted for the observed retrograde motions of the planets as successfully as did the system of homocentric spheres.

Retrograde motion was one problem; changing distance was another. Homocentric spheres cannot vary the distances of the planets from the earth. The epicycle system, however, accomplished this feat.

The epicycle hypothesis also had the great advantage of flexibility. If an astronomer found that the theory would not quite fit the observations, he could always try centering the deferent on a point other than the earth. Or he might try adding a smaller epicycle on the original epicycle. With enough circular motions, any motion can be reproduced.

Retrograde Motion in the Epicycle Hypothesis

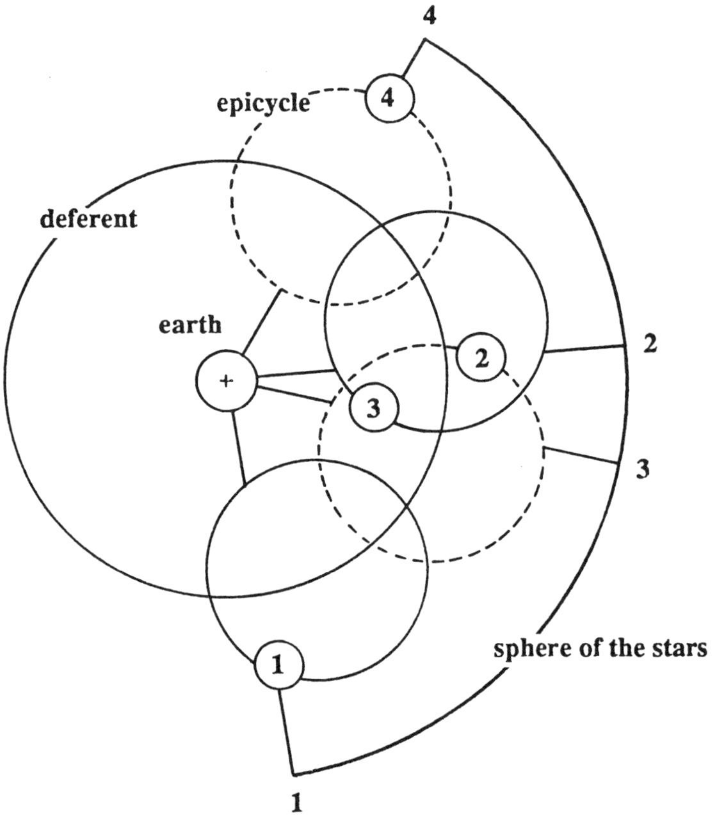

The deferent circle rotates around the earth. The center of the epicycle circle is attached to the deferent and carried around by it. The epicycle rotates around its center. The planet is attached to the epicycle and carried around by it. As seen from the earth, the planet appears to move against the sphere of the stars from 1 to 2, back to 3, and then resumes its forward motion toward 4.

Change of Distance in the Epicycle Hypothesis

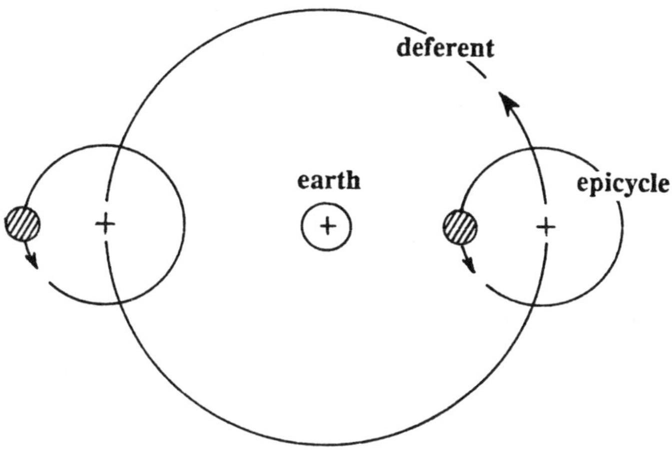

As the epicycle is carried from the left to the right side of the drawing, the planet is carried through 180 degrees by the rotation of the epicycle, bringing the planet closer to the earth.

However many the difficulties in Ptolemy's system, the system was a monumental achievement. It was unsurpassed for fourteen centuries. The vast data collected over several centuries at Alexandria and the cumulative development of a geometrical model by Apollonius, Hipparchus, and Ptolemy - each building on the achievements of his predecessors - was one of the great accomplishments of the Hellenistic world.

Scientific advances virtually ceased after Ptolemy, reflecting a general decline of society. Alexandria had already become a provincial city in the Roman Empire, and in the third century A.D. political crisis and almost incessant civil war in the Roman Empire disrupted internal order and economic production. Among the casualties were the Museum and Library.

The high point of development had been reached before 47 B.C., when Julius Caesar arrived by ship at Alexandria chasing Pompey. Pompey's head and signet ring were brought to the ship, but Caesar tarried in Alexandria. He was a scholar, and the Museum and Library probably were major lures; Cleopatra only smuggled herself into Alexandria and Caesar's presence after he landed. Caesar selected thousands of precious volumes from the Library to take back to Rome; they were lost in a fire that spread to the docks of Alexandria from the Alexandrian fleet destroyed by Caesar.

Caesar's was but the first of several imperial depredations after Rome gained ascendancy over Egypt, and the depredations grew more damaging during the tumult of the second and third centuries A.D. The major destruction of the Museum and Library occurred in the fourth century when, under emperor Constantine, Christianity triumphed in the Roman Empire and pagan institutions were subsequently destroyed. In 392 A.D. the last fellow of the Museum was murdered by a mob and the Library pillaged. Whatever may have remained was further damaged in the Arab conquest in the seventh century.

In addition to the decline and fall of Alexandria, which was followed shortly by the general decline and fall of the Roman Empire and Western civilization, there are several more specific reasons for the fourteen centuries of neglect of Ptolemy's astronomy, until Copernicus in the sixteenth century A.D.

First, the Greek science transmitted to the Latin West was that which had been summarized in Greek handbooks, which in turn were summarized further in Latin handbooks for readers who desired a smattering of science but nothing very profound. Ptolemy wrote the *Almagest* after the period of handbook transmission; consequently his book was known only by reputation in the West.

In the East, the emerging Arab civilization enjoyed a knowledge of Ptolemy's system. Arab scientists, however, were more interested in physics than mathematics, in Aristotle than Plato, and Ptolemy had limited his study to mathematics. When Greek science was first recovered in the Latin West, through Arab intermediaries, the preference for physics was also transferred. Not until the Humanistic demand of the fifteenth century for anything and everything Greek was Ptolemy avidly studied.

The scientific institutions and accomplishments of Alexandria were a product of the period of prosperity during the last three centuries B.C. And they were a victim of the general decline that followed. Frequent changes of rulers disrupted the continuity necessary for further advances. Changes of religion also brought disruptions, spreading throughout society, including that portion of society concerned with science.

The particular direction which the science of astronomy took was importantly influenced by problems internal to science, especially the phenomena of nature. Some major choices were arbitrary, though, including the focus upon finding a system of uniform circular motions with which to save the phenomena. The general goal of astronomy thus was as much or more a product of society and civilization as a model of the universe inherent in the phenomena. Furthermore, the level of scientific activity was and is dependent to a considerable extent on external societal factors.

Bibliography

On Ptolemy's and earlier planetary systems, see Thomas S. Kuhn's *The Copernican Revolution* and J. L. E. Dreyer's *History of the Planetary System from Thales to Kepler*, reprinted as *A History of Astronomy from Thales to Kepler.*

Internal inconsistencies in Ptolemy's planetary system and historical breaks in research in the Ptolemaic system are analyzed by Norriss S. Hetherington and Colin A. Ronan in their article "Ptolemy's *Almagest*: Fourteen Centuries of Neglect", published in the *Journal of the British Astronomical Association, volume 94, number 6* (October, 1984), 256-262.

On the city of Alexandria and the Museum and Library, see P. M. Fraser, *Ptolemaic Alexandria.*

The decline of Greek science is discussed in Ludwig Edelstein's admirable article "Recent Trends in the Interpretation of Ancient Science", published in the *Journal of the History of Ideas, volume 13* (1952), pages 573-604. The article is reprinted in Edelstein's *Ancient Medicine* and in P. Weiner and A. Noland, editors, *Roots of Scientific Thought.*

Excerpts from Ptolemy's *Almagest* are from the translation by R. Catesby Taliaferro, published in volume sixteen of *Great Books of the Western World.* Reprinted by permission from *Great Books of the Western World,* copyright 1952 by Encyclopedia Britannica, Inc.

The film *The Romans* (24 minutes) in the *Outline of History of Europe* series pictures the origins, growth, and collapse of the Roman Empire.

Epitaph for a Lost Civilization and Prolegomenon for the New

Twice in the history of man science has flourished.

Once was within Greek civilization, from roughly 400 B.C. to 150 A.D., and from Athens to the coast of Asia Minor and to the delta of the Nile River. There were institutions, private schools such as the Academy and the Lyceum, and public institutions, such as the Museum and the Library. There were individuals: Plato, Aristotle, Euclid, Archimedes, Ptolemy, and many others.

The West's second scientific revolution has its beginning fixed, somewhat arbitrarily, in 1543 A.D. This is the date of publication of Copernicus' *de Revolutionibus Orbium Caelestium*, of Vesalius' *de Fabrica*, and of a Latin edition of Apollonius' *Conic Sections*.

This second scientific flowering has now lasted over four hundred years. It is nearing the lifespan of the earlier Greek scientific revolution.

Not only the ancient glories of Greek science, but also the remarkable scientific achievements of megalithic Britons, of the Babylonians, of the Mayans, and of other civilizations, the ruins of some perhaps not yet discovered, stand as a symbol of what man can accomplish - and as a warning of the fragility of civilization. The pattern of the past suggests that our flourishing civilization is not immune to corruption and decay, that the impetus of original investigation and the advance of knowledge could again fail, and with it our civilization. Man may again face a period similar to that which earned the sobriquet "The Dark Ages".

An understanding of what happened and why it happened may help man avoid a repetition of his past. Understanding may help achieve a better future. The study of science and civilization, enjoyable for its own sake and uplifting to the soul, may also have something to contribute to the task of preserving and improving material aspects of our civilization.

Yet, paradoxically, the first and foremost lesson from Greek science is that progress then depended on the presence of a spirit of inquiry. Knowledge was valued for itself, without reference to potential utilitarian applications of science. The primary motivation behind the best of Greek philosophy was not how much money it would bring - Thales and the olive crop notwithstanding - but on how thought elevated the soul, on the sheer joy of knowing.

This is not said in advocation of the extreme position that the fruit of abstract science justifies the existence of the orchard of Greek people. Man does not live solely to exemplify some higher principle. Human life itself is sacred. But neither do people gain satisfaction solely from material goods. The Athenians constructing the Parthenon did so for the spirit's sake, and who can say with certainty that the effort would have been better put into preparation for war with Sparta? Man may not be able to eat beauty, or win battles with beauty, but neither can he live a beautiful life without beauty.

Nor can mankind expect a harvest of applied science year after year if the tree of knowledge is neglected. That is what the Greeks, and even more the Romans, did. Effort shifted from the increase of knowledge to the preservation of knowledge. Commentaries furnishing the results of Greek science were increasingly published, in place of new research continuing the aims and methods of Greek science. The spirit of inquiry faded. And as the commentators became less conscientiously critical, less careful to understand what they were summarizing for future generations, so even knowledge already obtained atrophied.

The early Greeks are not entirely blameless for this sorry state of affairs. At the other extreme, they had worshipped the moral value of knowledge and disdained its practical uses. Little if any understanding was achieved of the necessary balance between the pursuit of pure knowledge and its practical value for society, a balance needed if science is to be an integral part of civilization, supported by society in a manner that encourages a continuous tradition of progress. Greek science, with temporary exceptions such as the Museum and Library at Alexandria, usually consisted of little more than a few men committed to the pursuit of knowledge. There was virtually no institutional encouragement of the activity of science. The brightest minds were readily recruited away, in later years, into another service; Christianity became established institutionally in a way that science did not in the ancient world.

Democracy, such as it was at times, in Greece and Rome undoubtedly made more difficult an intelligent integration of science and society. It was not enough for some wealthy man to offer patronage in return for the glory of the resulting scientific accomplishment. An entire society, or at least a majority of those voting, had to appreciate both the spirit of inquiry and the potential utility of knowledge, when even the scientists themselves had been unable or disinclined to link knowledge and its application.

The initial steps in the scientific revolution of the sixteenth and seventeenth centuries A.D. were financed by dictators. The Medici family could wish for no more lasting monument than the satellites of Jupiter, called by Galileo, in search of patronage, the Medician stars. And much of America's scientific progress in the nineteenth century was funded by wealthy robber barons, in private universities. The French observer and writer Alexis de Tocqueville, after his travels in America in the 1830s, feared that a democracy would never support pure science. Nor did a general indifference in America toward basic science through much of the nineteenth century allay his fear.

Support for science and civilization has been democratized. Has an appreciation of the links between the two and the wisdom to preserve and advance science and civilization also been democratized? The stagnation of parts of our modern industrial economy is not assuring. There are, however, incredible successes suggesting a longer and happier future for our civilization.

Mankind today, on several continents if not uniformly over the face of the earth, enjoys the most extensive system of higher education ever developed. There is support for inquiry for its own sake, into fields as esoteric as relationships between science and civilization. And there is recognition of the failure of past civilizations and concern for the continuance of our civilization.

All these are encouraging signs of a better future.

Index